W0235322

Onscreen Chemistry
The Portrayal of Chemical Science in Film and TV

Onscreen Chemistry
The Portrayal of Chemical Science in Film and TV

By

John O'Donoghue
Trinity College Dublin, Ireland
Email: john.odonoghue@tcd.ie

ROYAL SOCIETY
OF **CHEMISTRY**

Print ISBN: 978-1-83767-214-1
PDF ISBN: 978-1-83767-435-0
EPUB ISBN: 978-1-83767-436-7

A catalogue record for this book is available from the British Library

© John O'Donoghue 2025

All rights reserved

Apart from fair dealing for the purposes of research for non-commercial purposes or for private study, criticism or review, as permitted under the Copyright, Designs and Patents Act 1988 and the Copyright and Related Rights Regulations 2003, this publication may not be reproduced, stored or transmitted, in any form or by any means, without the prior permission in writing of The Royal Society of Chemistry or the copyright owner, or in the case of reproduction in accordance with the terms of licences issued by the Copyright Licensing Agency in the UK, or in accordance with the terms of the licences issued by the appropriate Reproduction Rights Organization outside the UK. Enquiries concerning reproduction outside the terms stated here should be sent to The Royal Society of Chemistry at the address printed on this page.

Whilst this material has been produced with all due care, The Royal Society of Chemistry cannot be held responsible or liable for its accuracy and completeness, nor for any consequences arising from any errors or the use of the information contained in this publication. The publication of advertisements does not constitute any endorsement by The Royal Society of Chemistry or Authors of any products advertised. The views and opinions advanced by contributors do not necessarily reflect those of The Royal Society of Chemistry which shall not be liable for any resulting loss or damage arising as a result of reliance upon this material.

The Royal Society of Chemistry is a charity, registered in England and Wales, Number 207890, and a company incorporated in England by Royal Charter (Registered No. RC000524), registered office: Burlington House, Piccadilly, London W1J 0BA, UK, Telephone: +44 (0)20 7437 8656.

For further information see our website at www.rsc.org
For general enquiries, please contact books@rsc.org
For EU product safety enquiries, please email books@rsc.org or contact
Royal Society of Chemistry Worldwide (Germany) GmbH, Römischer Hof, Unter den Linden 10, 10117 Berlin.

Printed in the United Kingdom by CPI Group (UK) Ltd, Croydon, CR0 4YY, UK

For Maurice

Preface

I love movies. To me, they represent a unique method of story-telling in a dense but relatively compact package. The over-arching story can be extended in every direction with prequels, sequels, and even side stories. Growing up, I was very lucky that my small town in Ireland had an independent cinema that I could cycle to. I still remember the excitement when it expanded from a two-screen 'cineplex' into a three-screen 'multiplex'. Although it should be said that the third screen barely qualified as going to the cinema, some people probably had a similar-sized rear-projected screen in their house at the time.

The little tuck shop in our cinema was always packed full of every sort of treat you could imagine. It also had fresh popcorn, not the pre-made stuff that comes in large bags. The smell of popcorn would fill the entire building and waft out onto the street, especially on a hot summer evening. Of course, I don't want to give the impression that I am reminiscing through rose-tinted glasses, because the floors were usually sticky and the speakers in screen 2 crackled when the movie got too loud! Nonetheless, going to the cinema was a social experience. It was a place to go with your family, friends or on a date. For longer showings there was also an intermission where you could take a toilet break, get more snacks and chat about the movie so far. Although the cinema in my town survived numerous challenges

Onscreen Chemistry: The Portrayal of Chemical Science in Film and TV
By John O'Donoghue
© John O'Donoghue 2025
Published by the Royal Society of Chemistry, www.rsc.org

over the many years of its existence, it recently shut its doors for the final time.

I also grew up before the advent of streaming, so a TV series was a social event played out over many weeks and months. Every week we had to wait for the next episode at the same (Bat-) time, on the same (Bat-) channel. People would rearrange their day and weekly schedule to make sure they didn't miss an episode of a popular series. There was no 'plus one' option for TV channels, so the only way of watching an episode after it aired was by recording it on a VHS tape. However, missing an episode also had social implications, since every episode would be discussed and critiqued thoroughly during the intervening period at school, at work or during a night out. There was plenty of time to form opinions, so everyone was a critic!

The social side of watching movies and series still exists, but the conversation now revolves around the entire series rather than individual episodes. People usually recommend a series once they've finished streaming it, but discussions about the content can be hit and miss due to the asynchronous nature of streaming. To counter this, 'watch parties' have recently developed to synchronise content regardless of where people are consuming it. So, friends in different parts of the world can all watch something together at the same time, allowing for synchronous discussion. Also, the cinema is still popular for big blockbusters, proven by the 'Barbenheimer' cultural phenomenon during the summer of 2023. Admittedly, though, in recent years I haven't gone to the cinema as much as I would like.

But what about 'onscreen chemistry'? I grew up with 1980s and 1990s TV cartoons, many of which are mentioned in this book and some of which have chemistry references. But I can't claim that movies or TV influenced me to pursue a career in chemistry. My interest in science grew gradually over time, encouraged by the chemistry and physics sets I got from my parents. If I was influenced by anything onscreen it was most likely the mind-bending concepts of some sci-fi movies. To me, chemistry always felt like a practical science that was useful in the real world with lots of applications. After school I moved to the city of Cork in Ireland to study chemistry at university. It was there that I was lucky enough to experience the old Capitol Cinema before it shut down. Built in the 1940s, the Capitol cinema still retained much of its mid-20th century charm even by the time I got there. Complete with red curtains, it felt like a theatre experience and stood apart from the new, sleek, and modern cinemas.

This book has been over two decades in the making, since I first started collecting notes about chemistry in the movies during my time at university. This initial work led to a curiosity to find more, but it was during the COVID pandemic that the real work began. Encouraged by several people, I first presented my notes as an online public talk for Science Week in 2020. The talk gathered significant interest from scientists and non-scientists, leading to several demands for more talks at other festivals and events. More recently, I have adapted the original online talk for in-person events with a surprisingly large interest. To date I have given over a dozen talks about 'Chemistry in the Movies', adding new discussion points each time. The tagline for the talk was 'The Good, the Bad and the Ugly', to describe the different types of onscreen chemistry. This is also the title of one of my favourite movies, owing to the unique character development, gorgeous cinematography, and fascinating exploration of greed against a backdrop of the US civil war.

Writing for this book began when it became clear that there were significantly more onscreen chemistry examples than I had originally assumed. This book is a combination of my two favourite subjects from secondary school, namely English and Chemistry. So, thank you to all the teachers, educators, supervisors, mentors, and principal investigators who encouraged and inspired me throughout my career to date. Special thanks to Queens University Belfast and the Royal Society of Chemistry NI Local Section for inviting me to speak to a packed room on a rainy night for the NI Science Festival. Thanks also to the Royal Society of Chemistry Belgium Local Section for inviting me to talk at their annual awards ceremony. The popularity of those events hugely encouraged and inspired me during the writing of this book. Thanks also to Trinity College Dublin for access to all the references as well as to my colleagues for their encouragement and support. Similarly, thanks to the Royal Society of Chemistry for their support of this work.

This book has been a labour of love for about three years, and it would not have been possible without the unwavering support of my wife, Grace. Throughout the late evenings and weekends, early mornings, and lunchtimes, she gave me the space and time to scan through hundreds of movies and TV series. She also encouraged me, listened to my ramblings, and took interest in the numerous discoveries I made as I slowly weaved my way through the entire history of moving pictures. Thank you also to my friends, many of whom are probably sick of hearing me talk about this book and the movies described herein. Thanks also to my co-worker Socaí, who kept a vigilant guard over me while I researched, watched, and wrote through the late evenings. Thanks also to my family for their patience and support, please accept my apologies for all the events I missed while I was writing this book.

This book is dedicated to my brother Maurice. Together, we never missed an episode of the *Teenage Mutant Ninja Turtles* (1987–1996), *Batman: The Animated Series* (1992–1995), *A.J.'s Time Travellers* (1994–1995), *Earthworm Jim* (1995–1996) and *Pinky and the Brain* (1995–1998) among many others. It was like taking a journey back in time watching these again for the first time in decades to look for onscreen chemistry.

John O'Donoghue,
Trinity College Dublin, Ireland

Contents

Onscreen Chemistry: The Portrayal of Chemical Science in Film and TV
By John O'Donoghue
© John O'Donoghue 2025
Published by the Royal Society of Chemistry, www.rsc.org

SEASON 0

The Pilot

0.1 WHAT'S THE STORY?

This book will discuss the depictions of chemistry in visual media since the beginning of moving pictures, with a focus on influences in both directions. What I mean by that is the influence of the real world on depictions of chemistry and the influence of onscreen chemistry on the real world. The first is easy to see, with dozens of examples provided throughout this book. However, the second is much more difficult, so we must rely on nuances and the influence of early onscreen chemistry on later depictions. Overall, this book will mainly concentrate on depictions with explicit imagery or dialogue about chemistry, rather than those which imply there is chemistry involved. Although every care has been taken to include every known example, some may have slipped through the net. However, every theme and discussion point are supported by multiple examples throughout.

This book is divided into chapters called 'Seasons' which roughly follow each decade of moving pictures with some overlaps for the sake of effective storytelling. But what do we mean when we say 'story'? In 1968, the British literary scholar Barbra Hardy wrote "We dream in narrative, day-dream in narrative, remember, anticipate, hope, despair, believe, doubt, revise,

Onscreen Chemistry: The Portrayal of Chemical Science in Film and TV
By John O'Donoghue
© John O'Donoghue 2025
Published by the Royal Society of Chemistry, www.rsc.org

criticize, construct, gossip, learn, hate, and love by narrative".[1] In recent years, storytelling has also been identified as an important part of effective science communication.[2-5] But of course, storytelling has always been an innate part of human communication that has spanned our entire history. It transcends cultures, recalls history, builds relationships, and provides entertainment.

I'm from a town called Listowel, often referred to as the 'literary capital of Ireland' owing to the large number of published authors and storytellers from the surrounding areas. Some examples include John B. Keane (playwright, poet and novelist), Brendan Kennelly (poet, novelist and academic), Eamon Kelly (playwright, actor and Tony Award nominee), and Bryan McMahon (poet, playwright, academic and writer) among many others. Much of this rich storytelling heritage is honoured at the annual Writer's Week festival and in the town's Literary Museum, called the 'Kerry Writers Museum' or Seanchaí centre, as it is also known. *Seanchaí* is the Irish word for 'storytellers' and they were considered the definitive form of entertainment in ancient Ireland, holding the key to Irish folklore, myth, tradition, and legend.

So, what types of stories are there? and what is the story of onscreen chemistry? It is generally accepted that there are either five or six main types of story or narrative arcs that appear in novels, movies and other works.[6] The first is the very familiar 'Rags to Riches', also called 'the rise', and well-known examples of this include *Cinderella* (1950) and *Good Will Hunting* (1997) among many others. It should be pointed out of course that 'riches' don't always refer to financial gain; a character may be 'better off' romantically or emotionally at the end as well. Next is the opposite of this, called 'Riches to Rags' also called 'the downfall'. Examples include *Dracula* (1931) and *There Will Be Blood* (2007) among others where someone or thing is 'worse off' at the end, usually because they deserved it.

The next two are also opposites of each other with 'Recovery' describing a protagonist who might start at the top but experiences a fall and then rises again. The opposite of this is 'Pride and Fall', which follows a temporary rise before we eventually return to where we started. However, the best stories often contain a little bit of everything, often called 'The Journey', they contain a mixture of pitfalls and successes. The journey could start with someone or thing at the top, then fall, then rise again then fall again, or it could also do the opposite. These are subjective of course since the story of Cinderella could also be considered a journey because she returns to her 'wicked stepmother' at the stroke of midnight before she is eventually found by the prince as the one who fits the shoe. Nonetheless, it is very useful to keep the concept of story arcs in mind as we progress through the history of onscreen chemistry.

In recent years it seems that chemistry is undergoing a renaissance in terms of depictions in mainstream visual media, despite an apparent neglect in factual programming. Over the course of this book we will explore how modern depictions of onscreen chemistry have been influenced by the depictions that came before them, in terms of visuals, dialogue and stories. For instance, when did the colourful liquids in lab glassware start? What are the origins of the 'mad scientist' image? Has chemistry always been associated with explosions? Are chemists portrayed as good or bad? This book will trace the origins of modern chemistry depictions in addition to exploring the originality and evolution of some well-trodden concepts.

It turns out that there are many examples of onscreen chemistry across every decade throughout the entire history of moving pictures, increasing in number over time. In fact, there are a lot more than I anticipated at the outset of this project! As we progress through the decades, it will become very apparent that the depiction of chemistry in visual media has evolved over time. It will also become apparent that it has been influenced by the real world, in addition to the depictions that came before. So, the question at the heart of this book is: is art imitating chemistry or is chemistry imitating art?

0.2 THE SMELL OF CINEMA

The Ig Nobel Prize has been awarded annually since 1991, celebrating unusual or trivial achievements in scientific research. The name is a play on the well-known Nobel Prize, instead representing 'ignoble achievements'. It aims to "honour achievements that first make people laugh, and then make them think". The awards were created by Marc Abrahams and the Ig Nobels are presented by real Nobel laureates in an official ceremony at the Sanders Theatre at Harvard University. Although the prizes are satirical, they celebrate real research and often contain criticisms of real-world issues. In 2010, Andre Geim won the Nobel Prize in Physics with Konstantin Novoselov (for experiments with graphene) and in doing so became the first individual to win both a Nobel Prize and an Ig Nobel Prize.

Geim's Ig Nobel Prize was awarded in the year 2000 for using magnets to levitate a frog.[7] This work is now being used by Chinese researchers to develop an 'artificial Moon' facility that simulates low gravity using magnets.[8] But of course this isn't the first time that scientists have used frogs in their research. In 1780, the famous Italian physician, biologist and philosopher Luigi Galvani discovered that the muscles of dead frogs twitched when they were struck by an electric spark. Galvani's discoveries inspired the English author Mary Shelley to create Franken-stein's monster in her famous novel *Frankenstein; or, The Modern Prometheus*, first published in 1818 (Figure 0.1).[9] In Greek mythology, Prometheus gave fire to humanity in the form of

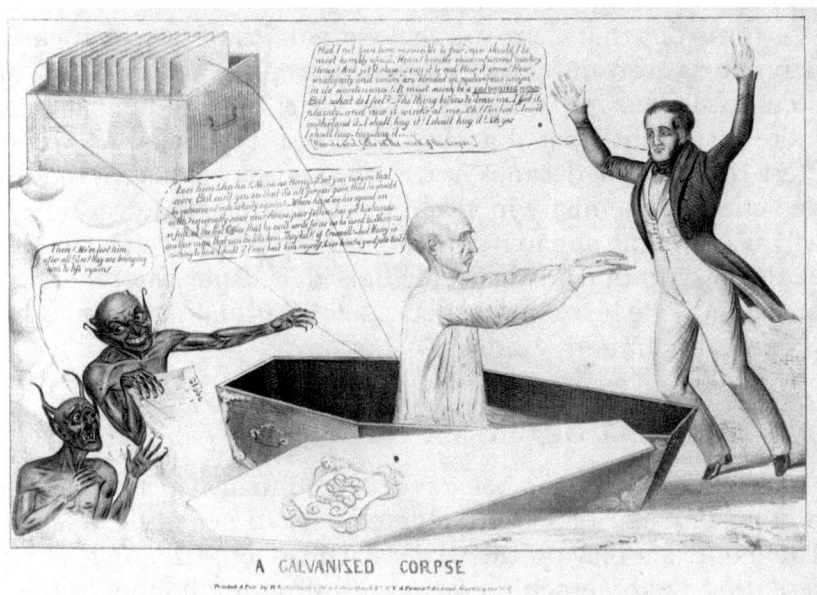

Figure 0.1   A 19th century cartoon depicting a corpse being brought back to life using 'galvanism'. Library of Congress/Science Photo Library.

technology and is also credited with the creation of humanity from clay.[10] Prometheus is referenced again in the title of the biography *American Prometheus: The Triumph and Tragedy of J. Robert Oppenheimer*, which was recently adapted for screen in *Oppenheimer* (2023).[11] It is now accepted that Mary Shelley's novel is referring to Galvani's attempt to create or reanimate life, while Oppenheimer gave the world fire in the form of the atomic bomb.

Also relevant to this discussion is the 2021 Ig Nobel Prize in Chemistry, which was awarded to Jórg Wicker *et al.* "for chemically analysing the air inside movie theatres, to test whether the odours produced by an audience reliably indicate the levels of violence, sex, antisocial behaviour, drug use, and profanity in the movie the audience is watching".[12] It explores the exhalation of volatile organic compounds (VOCs) and their connection to emotional responses. A cinema proved to be an ideal controlled environment for carrying out this analysis since it contains many people all watching and reacting to the same events on screen simultaneously. The researchers analysed the air from the ventilation system over 4 weeks using mass spectrometry, collecting VOCs from an impressive nine thousand people!

From changes in the CO_2 levels alone, they could determine when a movie was showing and when the theatre was empty, as well as when the cleaning staff were busy preparing the room for the next showing. While comparing the data from multiple showings of *The Hunger Games: Catching Fire* (2013), the CO_2 increased and decreased at the exact same time during every showing of the movie. The researchers then worked backwards using the movie timestamps and discovered that increases in CO_2 were directly associated with fight scenes in the movie. Researchers also found links with isoprene, which was associated with suspense and romance, while siloxanes were associated with injury. Ammonia and acetone were also linked with blood (violence), as well as formaldehyde with comedy.[12]

Because movies don't have specific labelling for every scene, only an overall age-related categorisation (PG 12, 15, 16, 18, R, *etc.*), the researchers needed to manually assign labels for scenes, *e.g.* action, romance, comedy, *etc.* Using their new system and the VOCs that were detected, they could predict forward and backward for specific types of scenes. Thus, the researchers created a potentially new method of categorising movies. They proposed that by measuring the VOCs during a test screening of a new movie, a more accurate categorisation could be established before it goes on to general release.

0.3   FACTUALLY NEGLECTED

Previous work in relation to cataloguing factual onscreen chemistry has found how little chemistry is highlighted.[13,14] *NOVA* (by PBS) is the most popular primetime science series on American television, which aims to demystify scientific and technological concepts. However, chemistry was found to be the least popular area chosen for broadcasting, behind all other areas of science.[13] Similar results were also reported for the Discovery Channel which found no chemistry programmes, but there were fifty-four about sharks.[15] There are also numerous recent programmes focusing on the broad theme of astrophysics such as *Through the Wormhole* (2010–2017) by the Science Channel and *The Universe* (2007–2015) on the History Channel among others.

However, chemistry overlaps with several areas of science by its central nature, so deciding whether something is chemistry-focussed or not can sometimes be a personal preference. Notwithstanding this, previous research found only six feature length PBS *NOVA* programmes that were clearly and primarily about chemistry, out of a total of approximately six hundred and ninety.[13] The chemistry features were *Linus Pauling: Crusading Scientist* (1977), *A Pill for the People* (1977), *Plague on Our Children* (1979), *Hidden Power of Plants* (1987), *Race to Catch a Buckyball* (1995), and *Forgotten Genius* (2007). However, during the research for this book, two recent multi-part series featuring chemistry were also found: *H2O: The Molecule that Made Us* (2020)[†] and *Beyond the Elements* (2021). Interestingly though, in both cases, these were filed under "Physics + Math" and "Nature" because there is no category for chemistry. The only other factual US chemistry themed TV series of note here is *The World*

[†] This title is not a typo: the 2 is not subscript in the chemical formula for water. The same is true when the BBC aired the same programme later.

of Chemistry (1990) hosted by the chemistry Nobel laureate Roald Hoffmann.[‡]

On the other side of the pond, the same was also found for the BBC's flagship science programme *Horizon*.[16] Even the most recent BBC *Horizon* special on vaccines only focussed on the genetic and biological aspects of their development, not the chemical. In fact, we need to go all the way back to the 1960s to find BBC *Horizon* specials which focus on chemistry in *The World of Buckminster Fuller* (1964–1965) and *Investigating Murder* (1967–1968).[16] Research into other UK factual science programming found that the majority of recent productions have focussed on broad biological themes such as *The Blue Planet* (2001), *Blue Planet II* (2017), *Planet Earth* (2006) and *Planet Earth II* (2016), not to be mistaken with the 1986 US PBS TV series also called *Planet Earth*.

However, there is one exception to note here, which is the BBC mini-series *Chemistry: A Volatile History* (2010) hosted by Jim Al-Khalili. This popular series was nominated for a British Academic Television Award and was later aired in the US as well. There are also some recent documentaries on streaming platforms focussing on the broad theme of radiation, which usually mention chemical elements, *e.g. Meltdown: Three Mile Island* (2022) and *The Days* (2023) about the Fukushima nuclear meltdown. Finally, worth a brief mention here also in terms of 'factual programming' are the ever-popular *Periodic Table of Videos* (2008 to present) featuring Martyn Poliakoff at the University of Nottingham. These are freely available through YouTube and their own website.[§]

Also worth a brief mention here is the British 'entertainment science series' *Brainiac: Science Abuse* (2003–2008). Hosted by Richard Hammond, it took a fast-paced and comedic approach to factual science and experimentation. Using a similar premise to *MythBusters* (2003–2016), experiments were carried out during each episode to verify whether commonly held conceptions were

---

[‡] *The World of Chemistry* with Roald Hoffmann (1990) is freely available online from various sources including YouTube https://www.youtube.com/@TheWorldofChemistry/videos.

[§] The Periodic Table of Videos is available through the University of Nottingham website https://www.periodicvideos.com or their dedicated YouTube channel https://www.youtube.com/user/periodicvideos.

true or false. For instance, is it possible to run across a pool of custard? Spoiler: the answer is yes. However, most often, the goal of the show was to create the most impressive explosions. Although chemistry did feature from time to time, the production team later admitted that some of the explosions were faked or "enhanced for visual effect".

A particularly famous chemistry-themed episode of *Brainiac: Science Abuse* (2003–2008) involves an experiment to illustrate periodic trends in the alkali metal series of chemical elements. It starts off with accurate demonstrations of metallic lithium, sodium, and potassium with water. They then move on to rubidium and caesium, describing them as "the king and queen of alkali metals". They start by dropping what appears to be rubidium into a water-filled bathtub, describing it as "letting off a hand-grenade in a bathtub". Similarly, caesium is described as letting off a "depth charge in a bathtub". However, it later transpired that neither of these reactions were spectacular, so the production team substituted the alkali metals for real explosives onscreen. A cable connected to an offscreen detonator can be observed on the side of the bathtub for a brief moment during the caesium experiment.[17]

An earlier and more successful attempt at these reactions was shown on British TV in the 1970s as part of the Open University series of programmes. Here, rubidium splatters around as soon as it hits the water's surface with some parts sinking and creating more violent explosions. Interestingly, caesium creates a small explosion and destroys the apparatus by blowing a hole in the side of the container.[18] However, neither of them are akin to a grenade or a depth charge as claimed by *Brainiac: Science Abuse* (2003–2008). Similar experiments with rubidium and caesium have also been repeated elsewhere with somewhat disappointing results. One theory for the lacklustre results for these alkali metals is related to the drop in the volume of hydrogen produced as we go down the periodic group.[17]

0.4   EXAMINING THE THEORIES

Numerous reasons for the dearth of factual onscreen chemistry are theorised. The first relates to the types of problems studied by chemists. Various media experts have argued that science is only as interesting as the questions it asks, such as: What is the origin of the universe? Can we find cures for AIDS or cancer? It is said that these are captivating questions that make good subjects for books, articles, and TV programmes.[13] However, it can easily be argued that chemistry also asks these questions, like "how do we prevent the planet from overheating?" and "how do we efficiently store green energy?", the latter of which is connected to the Nobel Prize in chemistry in 2019 in the form of lithium-ion batteries. So, this theory for the lack of factual onscreen chemistry is questionable.

The second theory behind why factual programming avoids chemistry is the fact that chemistry is very difficult to visualise since it occurs at an atomic and molecular level. This forms one-third of the well-known education concept 'Johnstone Triangle', or 'the chemistry triplet' as it is also known.[19] The chemistry triplet is widely used to explain chemical concepts through (1) the macroscopic (experiments that are visible to the human eye), (2) the symbolic (chemical equations and diagrams), and (3) the sub-microscopic (atoms and molecules). However, similar difficulties with unknown visuals have not held back the production of other abstract concepts such as black holes, as portrayed in *Event Horizon* (1997) and *Interstellar* (2014) among others. So, there is little evidence for this theory either.

The final theory is the belief by the media, as well as others, that chemists do not communicate information about their fields effectively. It is claimed that when chemists do have a discovery of great public interest, they are not very good at letting the world know about it. It has been claimed that "biologists and physicists may not be any more articulate than chemists, but they are more practiced when it comes to public relations and promoting their work".[13] However, this creates a chicken and egg situation, raising the question of how or why biologists and physicists are 'more practised'? Instead, this author proposes here that this may in fact be a theory contributing to reality, *i.e.* because chemists are so frequently told that nobody wants to hear about their work, they don't communicate it.

Chemists and journalists often relate a story in very different ways and there may be a fear by those in the media that the audience are afraid of chemicals, *i.e.* 'chemophobia'.[20] The American Council on Science and Health describe chemophobia as a fear arising from 'scare stories' and exaggerated claims about the dangers of synthetic substances in the media.[21,22] However, there is little evidence to support the idea that a widespread fear of chemicals exists among the public. In fact, the majority of respondents to public surveys felt that chemistry makes everyday life better.[14,22] Therefore the term 'chemophobia' may only be relevant to describe the media avoidance of chemistry topics and chemicals, potentially due to an exaggerated or false belief about public perceptions.[23]

Instead, it seems that chemistry suffers from a lack of connection to the real world around us. This results in people having few associations with it and instead defaulting to memories of school when asked.[24] The themes of 'dangerous chemicals' and 'school chemistry' will be discussed later in this book, so the contribution of onscreen chemistry to perceptions of chemistry in the real world may be more pronounced than previously thought. Although there may be a dearth of factual chemistry depictions, there are a plethora of examples of onscreen chemistry in movies and on TV from the beginning of moving pictures up to the present day.

0.5   BUILDING ON PREVIOUS WORK

There have been several keys works previously published in relation to chemistry depictions onscreen. First published in 2009, *ReAction! Chemistry in the Movies* by Mark Griep and Marjorie Mikasen is a fantastic exploration of chemistry in the movies. The authors take an in-depth look at the chemistry depicted in a wide range of movies and extrapolate the concepts they find into deeper discussions. As a result, they provide significant background to the chemistry mentioned on screen, including a historical timeline of the discovery and how it appears in the real world. The themes and genres from their work were later discussed and summarised into 'dark' and 'bright' sides for *Chemistry in Primetime and Online* from the US National Research Council of the National Academies.[13] Both of these works have been an invaluable resource for the current work, particularly for early movie examples.[25]

However, because they were published in 2009 and 2011, respectively, they do not include important genre-defining examples of onscreen chemistry such as *Breaking Bad* (2008–2013). Other work has looked at some chemistry on TV but doesn't compare similar movies whose depictions of chemistry demonstrate an evolution of the genre over time. Examples of this include *Erin Brockovich* (2000) compared to *Dark Waters* (2019), and *Apollo 13* (1995) compared to *The Martian* (2015).[26] These evolutions are an important part of the discussion and overall story about onscreen chemistry, and allows us to examine what led to these changes. There are also several important movies previously overlooked which provide important context and new perspectives to the discussion, *e.g. Superman III* (1983), *Darkman* (1990) and *Fight Club* (1999) among many others.

All these missing movies are included in this book and the story here is told using a historical timeline rather than structured by theme as done previously. This format allows for the full story of onscreen chemistry to be told for the first time, tracking the evolution of depictions from the silent era through to the first talkies and on to the technicolour dreamscape. Several examples of real-world events are also provided throughout the current work to provide context for the dominance of certain themes. Real-world events had a profound influence on movie themes, which also inspired and influenced the image of chemistry over time. As a result, this book will discuss how these themes evolved and how they may have affected the public image of chemistry in the real world, perhaps resulting in a feedback loop.

Other work that helped in the creation of this book includes the *Chemistry on the Screen* web project by Chris Magee. This provided some discussion about previously missed movies like *Fight Club* (1999), in addition to TV shows like *Lost* (2004–2010).[27] Both of these examples reference nitroglycerin (nitro-glycerine), a favourite substance for onscreen chemistry with examples provided throughout this book. Some of Chris's examples and chemistry are mentioned in this book in connection with similar depictions. There is also a previous use of the title *Onscreen Chemistry* by Jonathan Hare, which appeared as a series of articles for *InfoChem* and, later, *The Mole*, both published by the Royal Society of Chemistry's *Education in Chemistry*. Many of these articles provided useful details about the chemistry in movies like *Apollo 13* (1995) and *Dante's Peak* (1997), among others.[28,29] Jonathan's work also includes a review of the chemistry in *Breaking Bad* (2008–2013), which was invaluable during the latter part of this book.[30]

Away from written works, the podcast *Scientifically – The Men in White Coats* (2021) by Andrea Sella also provided a very useful and broad perspective about how scientists are depicted across various media. Starting with the image of Mary Shelley's Dr Victor Frankenstein, it explores the "deranged, driven researcher, who, unthinking of the consequences of his work, would unleash the angry powers of science". This podcast examines the origins of the 'mad scientist' image and discusses the public image of scientists with respect to classic movies. It is claimed that people usually imagine a male with white hair wearing a white coat when asked to describe a 'mad scientist'. However, as we will discover over the course of this book, this image is not actually due to any one depiction. Instead, this trope is probably an amalgamation of several depictions over time.

Unfortunately, the podcast doesn't provide context in the form of other fictional examples from the same era such as *The Love Test* (1935), which does not feature 'mad scientists'. It also doesn't mention several excellent biographies featuring real chemists with no mention of madness, *e.g. The Story of Louis Pasteur* (1935) and *Madame Curie* (1943). Although the podcast does explore the evolution of the 'mad scientist' image over time, it doesn't explore the possibility that the 'madness' is caused by the science itself due to some unforeseen side effect, *e.g. The Invisible Man* (1933). The concepts of 'side effects' and 'accidents' are intrinsically linked to onscreen chemistry and evolve slowly over time, as we will see in this book.

The podcast also focusses on the concept of scientists who work alone but it doesn't mention the most important example of this, Dr Jekyll, who creates an alternate personality called Mr Hyde. The famous story by Robert Louis Stevenson first published in 1886 will be discussed in detail throughout this book. The podcast also discusses the infamous *Man in the White Coat* (1951), where the hysteria of the protagonist is referred to as 'mad'. However, this may not be the full picture since there are several previous onscreen examples of persistent and/or obsessive scientists. As we will discuss in this book, persistence (or obsessiveness) is also portrayed as a requirement for real world success in *Dr Ehrlich's Magic Bullet* (1940) and *Madame Curie* (1943). Later we will see how even sensible scientists are referred to as 'mad' because their persistence and hard work is difficult for others to understand. We will also discuss how this trope evolves from obsession to workaholic in *Hollow Man* (2000).

However, the podcast does include an excellent discussion about how 'villain' scientists are depicted as trying to change the world, sometimes to prevent ecological disaster. The 'hero' often stops the scientist to maintain the status quo, raising the question of which side we are on. This is an extension of the common and persistent trope of scientists not considering all the consequences of their work, which is perfectly encapsulated in the famous quote from *Jurassic Park* (1993) by Jeff Goldblum's character "Your scientists were so preoccupied with whether or not they could, they didn't stop to think if they should". This theme has been discussed previously for other sciences[31] and will appear throughout this book, with several chemistry-based examples provided from nearly every decade.

Finally, there are also several research studies related to onscreen chemistry which assisted in many of the discussions contained in this book. Some explore specific examples like *Son of Frankenstein* (1939)[32] and *Apollo 13* (1995)[33] while others examine a range of movies and themes,[34] all of which will be discussed here within a wider context. There are also works which look at specific concepts such as the fictional element vibranium,[35] the two sides of chemicals,[36] as well as aliens and extra-terrestrial minerals,[37] all of which will be explored and extracted further with more recent examples.

There are also some excellent previous works looking at science on screen in a general sense, this book will specifically focus on chemistry and chemists onscreen.[38,39] It will reexamine prominent themes with new perspectives and context, especially in relation to recent depictions like *Lesson in Chemistry* (2023). Many famous movies and serials have also been remade in recent years, offering us a unique opportunity to compare the evolution of onscreen chemistry for the same story. Also, all previous work has revolved around English-language depictions of chemistry, here we will also look at a selection of onscreen chemistry from around the world. Overall, *Onscreen Chemistry: The Portrayal of Chemical Science in Film and TV* will build on previous work and tread new ground in relation to movies, TV, miniseries, and animations. Herein is a comprehensive discussion which spans the entire history of moving pictures in every format.

0.6   OTHER ONSCREEN CHEMISTRY

Before we dive into the main discussion about onscreen chemistry in movies and on TV, it's worth briefly mentioning a few other types of onscreen chemistry here. First are TV advertisements, which sometimes feature labs, glassware, and scientists in lab coats. There is a specific example of this for a well-known anti-baldness shampoo, which features scientists in lab coats in a lab explaining their product. However, most ads also feature marketing hype such as "active oxygen bubbles" in certain brands of toothpaste. There are also several examples of famous people emphasising the science like Jennifer Anniston telling us "here comes the science bit, concentrate" to promote hair products and Eva Longoria enunciating "hyaluronic acid" for skincare. Surfactants are also mentioned frequently in relation to cleaning products, and the level of strength for most products is strangely defined as 'clinical' or 'industrial', depending on the application, *e.g.* industrial strength adhesive.

There are also several examples of onscreen chemistry in music videos, which depict similar scenes to those seen in movies and TV from the same era, reflecting the evolution of onscreen chemistry. First, we have Eurythmics – *Missionary Man* (1986), which opens in a lab with glassware, boiling liquids, and lots of tubing. A snake also slithers between the glassware as we are also shown some random electronics and gauges. The 'scientist' (Dave Stewart) pours a liquid into the apparatus which produces what appears to be a thick condensate of a similar appearance to dry ice (CO_2). The lead singer (Annie Lennox) is formed from some kind of wax in one of the flasks, like the artificial skin seen later in *Darkman* (1990). There are several other images of lab glassware filled with colourful liquids throughout the music video. The use of colourful liquids and tubing is a common trope during this era which will gain greater prominence towards the second half of this book.

We then have Semisonic – *Chemistry* (2001), which features an elaborate glassware setup behind the band as they draw comparisons between romantic chemistry and the scientific discipline. For instance, they call dating and courtship "conducting experiments". The glassware shown includes condensers, round-bottomed flasks, and burettes, all filled with colourful liquids and joined together with tubing in a nonsensical fashion. There's also a periodic table on the wall near the glassware and a white mouse in a cage, imagery that will be encountered throughout this book. Near the end of the video, we also get some close-up shots of the glassware, showing condensate pouring out of the flasks. This is again most likely dry ice (CO_2) since the liquids are also bubbling without an apparent heat source.

Next, we have Katy Perry – *Chained to the Rhythm* (2017) which takes place in a 1950s futuristic theme park. Briefly in the background we can see *The Atomium*, which is the real centrepiece of the 1958 Brussels World's Fair (Expo 58) that can still be seen and visited today. The nine eighteen-metre-diameter stainless steel-clad spheres are connected in the shape of a unit cell of an iron crystal magnified 165 billion times. As she enters the park, Katy is seduced by the delights and distractions of the modern world. But she slowly realises that something is wrong with the American dream, the origins of which is often traced back to the 1950s. This era is also referred to as the atomic age, where the technological possibilities and scientific discoveries drove American optimism about the future.[31]

This music video also includes a scene at a fossil fuel station called 'Inferno H_2O' which satirically links our dependence on oil to water, which is essential for life. All the fuel pumps are topped with large fishbowls filled with a flaming blue liquid. The flaming fuel is poured into a beaker and then served to customers in cars like burgers at a 1950s American diner. A character drinks the flaming blue liquid from a large beaker and then blows out flames. However, the 'Inferno H_2O' scene is not just referring to our dependency on oil, but also to the looming crisis related to the world's fresh water supply. It also seems to link chemistry to the fossil fuel industry and/or the freshwater crisis by showing everyone drinking from beakers.

Similarly, we also have poisonous chemistry and water in the music video for 21 Pilots – *Chlorine* (2018) which features large drums of blue liquid, a swimming pool, and a large metal reaction vessel. The lyrics include "sipping on straight chlorine, this beat is a chemical... the moment is medical". Elemental chlorine is a gas and will be discussed at several stages during this book. But since the video features liquids, we can assume they mean sodium hypochlorite or hypochlorous acid, which is the disinfectant normally added to swimming pools to kill bacteria. Sodium hypochlorite forms hypochlorous acid when added to pools, but it is yellow in colour. Hypochlorous acid can be obtained in large blue containers, but the liquid is colourless.

Although hypochlorous acid is safe to consume in diluted form and generally considered non-toxic, the concentrated form is corrosive and oxidising. Sodium hypochlorite, on the other hand, can lead to severe poisoning; in fact, even breathing the fumes can be poisonous. It also causes skin irritation, vomiting and corrosive injury to the gastrointestinal tract. So, although the colour is incorrect, this is the substance they are most likely referring to in the song since the lyric is "venom on my tongue... poisonous vibration". We will return to the duality of chlorine during the final discussion of this book.

Other brief examples include the video for Interpol – *Rest My Chemistry* (2007) which features some abstract images of atoms and molecules moving in different patterns. Despite the name, the video for Blur – *Chemical World* (1993) doesn't really include any onscreen chemistry. The same can be said in relation to the video for Fatboy Slim – *Better Living Through Chemistry* (1996) and The Chemical Brothers – *C-h-e-m-i-c-a-l* (2015). However, it should be noted here that the title *Better Living Through Chemistry* is referring to recreational drugs, which will be discussed in more detail later. There are also plenty of songs with titles that reference chemical elements, but they do not feature onscreen chemistry in their music video, for example Jean Michel Jarre – *Oxygene* (1976), Nirvana – *Lithium* (1991), Doves – *The Sulphur Man* (2002), and David Guetta ft. Sia – *Titanium* (2011), among others.

However, more recently the video for The Kid Laroi – *Wrong* (2020) features a brief scene in a school lab where two students in lab coats and safety goggles use a measuring cylinder and a conical flask to pour liquid into a 'volcano'. This is most likely the well-known vinegar and baking soda 'volcano eruption', which is a common school science demonstration of acids and bases. Some laboratory glassware also feature very briefly in relation to drugs/narcotics in the music video for Victony – *Soweto* (2022). Finally, worth a brief mention is Bell X1 – *One Stringed Harp* (2009), which contains the lyric "Like Wile E. Coyote/As if the fall wasn't enough/Those bastards from Acme/They got more nasty stuff". There's no music video for this song, but we will discuss the infamous Acme corporation later, especially in relation to *Who Framed Roger Rabbit* (1988). Acme also represents the comical peak of the 'evil industry' theme which will be discussed throughout this book.

ABBREVIATIONS

PBS	Public Broadcasting Service, US
BBC	British Broadcasting Corporation, UK
WW1	First World War
WW2	Second World War
CGI	Computer-Generated Imagery

GLOSSARY OF LAB TERMS

Beaker – A cylindrical container with a flat bottom. They also contain a 'beak', or a spout to aid pouring, possibly from the Ancient Greek word βικος (*bikos*), meaning a pot, bowl or drinking vessel. They usually contain graduations for roughly measuring liquid volume. They are available in several sizes and shapes. Tall beakers are also known as Berzelius beakers or measuring beakers, and in older examples were tapered; named after the Swedish chemist Jöns Jakob Berzelius.

Pipette – A laboratory tool for measuring and transporting precise volumes of liquid. There are several designs ranging from micropipettes, serological pipettes and Pasteur pipettes, which are named after the famous chemist Louis Pasteur.

Graduated cylinder – Also known as a measuring cylinder, it is a piece of lab glassware used for measuring accurate volumes of liquid. Due to the shape, liquids can form a curved meniscus on the surface, so they are designed to be read from the bottom of the meniscus.

Round-bottomed flask – A spherical flask with one or more inlet/outlet connections and made from heat-resistant borosilicate glass. They come in different sizes and are used for a variety of common chemical process such as distillation, reflux, heating, boiling, and more. Also sometimes known as a boiling flasks, but these usually have long necks.

Florence flask – A rounded flask with a flat bottom and long neck, usually used as a container for storing liquids for short-term use. It is also designed for uniform heating and ease of swirling. It is named after the Italian city of Florence.

Separation flask – Also known as a separatory flask or funnel, it is a piece of glassware used for liquid–liquid extractions and for separating mixtures into two solvent phases of difference densities.

Conical flask – Also known as an Erlenmeyer flask, it is conical in shape with a cylindrical neck and flat bottom. Its shape allows for swirling liquids with no spillage. It is named after the German chemist Emil Erlenmeyer. It can also be modified into a Buchner flask by the addition of a side arm for use with vacuum filtration.

Condenser – Officially known as a Liebig condenser or straight condenser, it is a long glass tube surround by another tube which acts as a water jacket. Using tubing, water is pumped through the water jacket to cool vapours inside the inner tube without mixing. They are usually used for distillation or reflux. There are various types with different configurations for different applications.

Test-tube – Also known as a culture tube or sample tube, it is a long cylindrical tube with one end closed, intended for general chemical work. They can withstand high temperatures and are generally used for handling chemicals, reactions, and qualitative assays.

Volumetric flask – A pear-shaped flask with a stopper used for measuring precise volumes at a certain temperature. They can also be used to prepare precise dilutions and preparations of standard solutions.

Fume hood – A closed or semi-closed local exhaust ventilation device. It is designed to prevent users from being exposed to hazardous fumes, vapours or dust.

Centrifuge – A piece of laboratory equipment that spins liquid samples at a high speed, usually driven by a motor. They can separate samples, particularly those containing sediment.

Microscope – From the Greek words *mikros* meaning 'small' and *skopeo* meaning 'inspect' or 'examine'. It is a laboratory instrument used to examine objects that are too small to see in detail by the naked eye.

Rotary evaporator – A device used for the efficient and gentle removal of solvents from samples by evaporation under low pressure.

Bunsen burner – Named after the chemist Robert Bunsen, it is an air–gas burner which produces a single open gas flame. The heat of the flame can be adjusted through the amount of oxygen allowed into the device. It is used for heating samples, sterilization and combustion.

REFERENCES

1. B. Hardy and D. Jervolino, Towards a Poetics of Fiction: 3) An Approach through Narrative, in Novel: A Forum on Fiction, *JSTOR*, 1968, **2**, 5–14.
2. S. N. Collins, The Importance of Storytelling in Chemical Education, *Nat. Chem.*, 2021, **13**(1), 1–2.
3. D. H. Torres and D. E. Pruim, Scientific Storytelling: A Narrative Strategy for Scientific Communicators, *Commun. Teach.*, 2019, **33**(2), 107–111.
4. S. Martinez-Conde and S. L. Macknik, Finding the Plot in Science Storytelling in Hopes of Enhancing Science Communication, *Proc. Natl. Acad. Sci. U. S. A.*, 2017, **114**(31), 8127–8129.
5. Y. Hadzigeorgiou, Humanizing the Teaching of Physics through Storytelling: The Case of Current Electricity, *Phys. Educ.*, 2006, **41**(1), 42–46.
6. A. LaFrance, The Six Main Arcs in Storytelling, as Identified by an A.I., The Atlantic, https://www.theatlantic.com/

technology/archive/2016/07/the-six-main-arcs-in-storytelling-identified-by-a-computer/490733/.

7. M. V. Berry and A. K. Geim, Of Flying Frogs and Levitrons, *Eur. J. Phys.*, 1997, **18**(4), 307–313.

8. S. Chen, China Has Built an Artificial Moon That Simulates Low-Gravity Conditions on Earth, *South China Morning Post*, 2022. https://www.scmp.com/news/china/science/article/316 2972/china-has-built-artificial-moon-simulates-low-gravity-conditions.

9. E. Blakemore, How Twitching Frog Legs Helped Inspire 'Frankenstein', Smithsonian Magazine, https://www. smithsonianmag.com/smart-news/how-twitching-frog-legs-helped-inspire-frankenstein-180957457/.

10. J. Weiner, B. E. Stevens and B. M. Rogers, *Frankenstein and Its Classics: The Modern Prometheus from Antiquity to Science Fiction*, Bloomsbury Studies in Classical Reception, London, 2018.

11. K. Bird and M. J. Sherwin, *American Prometheus: The Triumph and Tragedy of J. Robert Oppenheimer*, A.A. Knopf, New York, 2005.

12. C. Stönner, A. Edtbauer, B. Derstroff, E. Bourtsoukidis, T. Klüpfel, J. Wicker and J. Williams, Proof of Concept Study: Testing Human Volatile Organic Compounds as Tools for Age Classification of Films, *PLoS One*, 2018, **13**(10), 1–14.

13. T. Masciangioli, *Chemistry in Primetime and Online: Communicating Chemistry in Informal Environments: Workshop Summary*, The National Academies Press, 2011.

14. National Academies of Sciences, Engineering, and Medicine, *Effective Chemistry Communication in Informal Environments*, The National Academies Press, Washington DC, 2016.

15. Discovery Networks International, Discovery Channel UK, 2023, https://www.discoveryuk.com/series/.

16. The BBC, Horizon, 2023, https://www.bbc.co.uk/programmes/b006mgxf.

17. D. Fleming, Alkali metals – the camera lies. Education in Chemistry, https://edu.rsc.org/analysis/alkali-metals-the-camera-lies/2010008.article.

18. OpenLearn, *Alkali Metals*, The Open University, https://www. open.edu/openlearn/science-maths-technology/science/chemistry/alkali-metals.

19. A. H. Johnstone, You Can't Get There from Here, *J. Chem. Educ.*, 2010, **87**(1), 22–29.
20. K. Sanderson, What are you afraid of?, Chemistry World, https://www.chemistryworld.com/features/what-are-you-afraid-of/6732.article.
21. J. Entine, *Scared to Death: How Chemophobia Threatens Public Health*, American Council on Science and Health, 2011.
22. E. Fu, A. Fitzpatrick, C. Connors, D. Clay, B. Toombs, A. Busby and C. O'Driscoll, Public Attitudes to Chemistry, *R. Soc. Chem.*, 2015, ch. 4, pp. 47–54.
23. D. Blum, *Chemical-free nonsense*, Los Angeles Times, https://www.latimes.com/opinion/la-xpm-2012-jan-22-la-oe-blum-chemicals-20120122-story.html.
24. C. Ceci, Take Concepts of Chemistry out of the Classroom, *Nature*, 2015, **522**(7554), 7.
25. M. A. Griep and M. L. Mikasen, *ReAction!: Chemistry in the Movies*, Oxford University Press, New York, 2009.
26. S. Perkowitz, D. J. Nelson, K. R. Grazier and J. Paglia, *Hollywood Chemistry: When Science Met Entertainment: 1139 (ACS Symposium Series)*, OUP USA, 2014.
27. C. Magee, *Chemistry On The Screen*, University of Bristol, https://www.chm.bris.ac.uk/webprojects2006/Macgee/Web Project/home_page.htm.
28. J. Hare, *Apollo 13 – lithium hydroxide saves the day*, Education in Chemistry, https://edu.rsc.org/feature/apollo-13-lithium-hydroxide-saves-the-day-/3007380.article.
29. J. Hare, Soap: Can you make it with body fat and is there an explosive spin-off?, Education in Chemistry, https://edu.rsc.org/opinion/soap-can-you-make-it-with-body-fat-and-is-there-an-explosive-spin-off/2021319.article.
30. J. Hare, *The Science behind Breaking Bad*, Education in Chemistry, https://edu.rsc.org/analysis/the-science-behind-breaking-bad/2500089.article.
31. R. Jones, "Why Can't You Scientists Leave Things Alone?" Science Questioned in British Films of the Post-War Period (1945–1970), *Public Underst. Sci.*, 2001, **10**(4), 365–382.
32. B. Ruekberg, Another Useful Film Clip: Scientific Methodology of the Frankenstein Monster, *J. Chem. Educ.*, 2021, **98**(12), 4101–4103.

33. J. G. Goll and B. J. Woods, Teaching Chemistry Using the Movie Apollo 13, *J. Chem. Educ.*, 1999, **76**(2–4), 506–508.

34. N. C. Thomas, Using Classic Movie Chemistry Scenes to Introduce Classroom Activities, *J. Chem. Educ.*, 2021, **98**(5), 1814–1817.

35. S. Collins, T. Steele and M. Nelson, Storytelling as Pedagogy: The Power of Chemistry Stories as a Tool for Classroom Engagement, *J. Chem. Educ.*, 2023, **100**(7), 2664–2672.

36. D. J. Wink, "Almost like Weighing Someone's Soul": Chemistry in Contemporary Film, *J. Chem. Educ.*, 2001, **78**(4), 481–483.

37. M. A. Griep and M. L. Mikasen, Close Encounters with Creative Chemical Thinking: An Outreach Presentation Using Movie Clips about the Elemental Composition of Aliens and Extraterrestrial Minerals, *Educ. Quim.*, 2016, **27**(2), 154–162.

38. C. Frayling, *Mad, Bad and Dangerous? The Scientist and the Cinema*, Reaktion Books, London, 2005.

39. D. A. Kirby, *Lab Coats in Hollywood: Science, Scientists and Cinema*, The MIT Press, London, 2011.

First Principles

1.1 THE CHEMIST'S PHOTO LAB

The story of onscreen chemistry begins with the birth of photography in the 19th century, which was intrinsically linked to the advancement of science and experimentation at the time. Photography generally consists of two parts: a device (camera) and a method of storing the image (photograph). One of the earliest examples of a camera was the *camera obscura* from Latin meaning 'dark chamber', which consisted of a box with a small hole in one side. Light from an external scene passed through the hole onto a clear surface like a canvas or a sheet of paper, usually inverted (upside-down) and reversed (left-to-right).

To capture the image permanently for later viewing, the photographer (from Greek meaning 'drawing with light') needed to quickly trace over the image using drawing implements like pencils and brushes. Speed was important since clouds or other obstructions made it difficult to capture a scene accurately if the process took too long. However, the manual nature of this process also meant that even the most accurate photographer frequently added their own style and interpretation of the scene. The honesty and societal standing of the artist was therefore of great importance, since it reflected the accuracy of the scene being captured.

Onscreen Chemistry: The Portrayal of Chemical Science in Film and TV
By John O'Donoghue
© John O'Donoghue 2025
Published by the Royal Society of Chemistry, www.rsc.org

This method of manual image interpretation was also used to capture scientific discovery. In 1857, the Irish microscopist Mary Ward published a popular book called *Sketches with the Microscope* which contained incredibly detailed hand-drawn images of what she observed under her microscope. These images of insects and plants were some of the most detailed available at the time, and several prominent scientists enlisted her skills to produce images of their experiments and apparatus. The book was reprinted eight times between 1858 and 1880 under the subtitle *A world of wonders revealed by the microscope*, such was the incredible interest in her work.[1]

It is generally accepted that the earliest photograph was produced in France by Joseph Nicéphore Niépce in 1827. He called his photographic creations *héliographie*, or 'sun writing'. The development of the *héliographie* plate involved many years of experimentation with a variety of chemical combinations, with early experiments consisting of paper sheets coated in light-sensitive silver salts. Some of these experiments were successful in producing an image in a darkened room, but once the image was exposed to daylight, it would vanish as the silver salt continued to blacken. However, this early success inspired Niépce to research other compounds that were bleached by light instead of blackened. Eventually he found bitumen resin to be selectively light sensitive and somewhat resistant to acid, so a metal plate could be etched with acid where the bitumen had been removed by light.

After experiments with copper and limestone plates, pewter was found to provide the clearest image due to its reflective surface. Pewter is a malleable alloy that has been used to produce household and other items since ancient times, first used by the Egyptians and later the Romans. It consists mostly of tin (85–99%) with small amounts of copper, antimony, bismuth as well as silver to increase its hardness and durability. Some older forms of pewter also contained lead, but these were phased out due to lead toxicity on the human body which was well-known since ancient times. This fact will be discussed again later in relation to the use of leaded petrol/gasoline in the movie *One Man* (1977).

To make the heliograph, Niépce dissolved light-sensitive bitumen in oil of lavender and applied a thin coating over a polished pewter plate. He then inserted the plate into a *camera obscura* and positioned it near a window in his second-story laboratory. After several days of exposure to sunlight, the plate produced a grainy image of the courtyard, outbuildings, and trees outside which is now known as *View from the Window at Le Gras*. This direct reproduction of an image represented a significant leap forward since it was the first true representation of a scene, without human input or interpretation.[2]

Niépce's photographic process was improved further by another French chemist named Louis Daguerre, who brought photography to a worldwide audience in 1839. Instead of pewter, Daguerre used a polished sheet of silver-plated copper treated with silver iodide fumes to make the surface light sensitive. Postprocessing involved treatment with mercury vapour and sealing the result in a protective glass enclosure (Figure 1.1). Further experiments in the 1850s eventually sped up plate development significantly. In particular, the 'wet collodion' process became

Figure 1.1 This Daguerreotype photo from 1838 is believed to be the earliest photograph showing a living person. This is a busy street, but only a boot polisher and his customer stayed in one place long enough to be visible due to the time needed for the exposure. Science Source/Science Photo Library.

popular, which involved the use of an iodide-impregnated cellulose nitrate coating on a glass plate immersed in a solution of silver nitrate to form silver iodide.

The development of photography during the 19th century had a profound influence on the art world, since, for the first time in history, a painter was no longer required to produce a portrait or to capture a scene. Many artists turned to photography to embrace the new technology, which was then in high demand. However, others felt freed by the technology. Artists like Monet, Renoir, Degas, and later Van Gogh, instead decided to experiment with techniques that were different from what a camera could capture by playing with light, colour and brushstrokes. The influence of photography and the evolution in painting styles is particularly evident from Monet's work. In the 1850s and 1860s much of his work consisted of highly detailed scenes and portraits, whereas the well-known hazy and expressive style of impressionism eventually dominated the remainder of his life.

In 2007, the world's oldest surviving photographic lab from this era was re-discovered in France, complete with all the chemicals and apparatus used in the darkroom to prepare and develop photographic plates. It originally belonged to Joseph Fortuné Petiot-Groffier and the receipts found for the 450 flasks date the chemicals to 1840–1841, providing a fascinating insight into the early development of photography and its link to chemistry.[2]

1.2 LIGHTS, CAMERA, CHEMISTRY

Further developments of photographic plates and their commercial production laid the foundation for the immediate development of motion pictures in the late 19th century. The first motion pictures were created using optical illusions with simple devices made from paper and string, commonly known as thaumatropes. This invention is commonly credited to the British physician John Ayron Paris and became a very popular toy in Victorian England. A common example of this optical effect is the illusion of a bird escaping from a cage with one side of paper containing an image of a bird, while the other side has a cage. When the strings are twirled quickly, the illusion shows the bird trapped inside the cage and then freed again, using only two 'frames' or 'scenes'.

Another important early version of moving images was created in 1878 by Eadweard Muybrudge using chronophotography, which involved the combination of still photos to capture animal locomotion. He used multiple cameras and dozens of still photographs to capture all the positions of a galloping horse. He then combined the still images into 'cabinet cards' and later into a circular disk format for his zoopraxiscope device (like a movie projector) to create a short film called *Horse in Motion* (1878). In doing so, the true nature of animal locomotion was captured for the first time, which differed significantly to the four outstretched legs usually portrayed in paintings (Figures 1.2 and 1.3). So, like still photos, motion pictures had a significant influence on the scientific study of the world around us by capturing 'true' and 'accurate' depictions of movement for the first time.

However, to create longer films with hundreds of scenes, the photographic process needed significant advancements.

Figure 1.2 *The Epsom Derby*, painting by Théodore Géricault from 1821 showing the horses with all their legs outstretched. © Peter Horree/Alamy Stock Photo.

Figure 1.3   Horse at the amble, from *Animals in motion. An electro-photo-graphic Investigation of Consecutive Phases of Muscular Action*, by Eadweard Muybridge (1885). Universal History Archive/Science Photo Library.

Photographic plates made from pewter and other hardened materials required hundreds of rigid plates for each scene, as well as an elaborate contraption to move them in the correct sequence. A French inventor named Louis Le Prince did achieve such a mechanical device, which produced one of the first moving picture sequences using rigid photographic plates. *Roundhay Garden Scene* (1888) was recorded in Leeds, England, where Le Prince worked as a photographer and painter. The scene is approximately 2 seconds long and shows some of Le Prince's family walking around a garden. Le Prince's complicated system involved the use of coated glass slides dropping down a tube in front of a lens and then onto a belt which spiralled its way back to the top to continue the projection. He patented his system and was on route to New York for a public unveiling when he disappeared from a train in 1890. His body and luggage were never found, and his family openly suspected foul play in his disappearance.[3]

Instead, it was the invention of flexible polymers that directly led to the first substantial production of moving pictures.

Natural rubber or latex was one of the first polymers to find commercial success in the mid-1800s. The process of 'thermo-setting' a polymer causes it to be irreversibly hardened or 'cured', and for rubber this is known as 'vulcanization'. This process forms cross-links between the long rubber molecules to achieve better elasticity, residence, tensile strength, hardness, and weather resistance. Discovered by the American Charles Goodyear in 1839, he patented the chemical process of removing sulfur from rubber and then heating it to about 140–160 °C. However, although rubber is flexible enough to be used for strips of film, it is not translucent, thereby preventing any form of light projection.

A few years later, in 1845, the Swiss-German chemist Christian Friedrich Schönbein accidentally discovered nitrocellulose. The story goes that he spilled a mixture of nitric acid and sulfuric acid in his kitchen while carrying out some experiments, despite objections from his wife for using her kitchen as a lab. He then used his wife's cotton apron to mop up the spill, which spontaneously ignited as it was drying over the stove. All of this must have been much to his wife's disgust. However, Schönbein had accidentally converted the cellulose of the apron into nitrocellulose, with the nitro groups (added from the nitric acid) serving as an internal source of oxygen.[4]

Later this discovery would find extensive military use since, unlike ordinary gunpower, nitrocellulose is 'smokeless' when it ignites. A short time later it featured in numerous works by the famous 19th century French novelist Jules Verne who fictionalised its use as a rocket propellent in *De la Terre á la Lune, trajet direct en 97 heures 20 minutes* (*From the Earth to the Moon: A Direct Route in 97 Hours, 20 Minutes*) published in 1867.[†] Interestingly, a reference to this work appears in the American science fiction movie *Back to the Future: Part III* (1990). After mistakenly travelling back to 1885, the time-travelling Doc Brown (Christopher Lloyd) quotes from this novel in response to Clara Clayton (Mary Steenburgen) asking if mankind will ever "travel to the Moon the

---

[†] When the roller coaster Space Mountain opened in Disneyland Paris in 1995, it was named *Space Mountain: De la Terre à la Lune*, based loosely on this novel. The attraction's exterior was designed using a Verne-era influence with rivet and boiler plate effect and the Columbiad, which recoils with a bang and produces smoke, giving riders the perception of being shot into space. It was refurbished in 2005 and renamed Space Mountain: Mission 2 and no longer featured elements of the original storyline from the novel. It was re-opened again in 2017 with a *Star Wars* theme.

way we travel across the country on trains". We will return to Doc Brown and Marty later in this book, particularly in relation to the image of 'mad scientists'.

Both Jules Verne and the British novelist H. G. Wells published some of their most famous novels around the same time that motion pictures were being developed, cementing the link between science fiction and cinema from the very beginning. Interestingly, both authors frequently used real scientific facts in their writing. It was only when the scientist or another character bended or exaggerated one of those facts that we get the premise for many of their famous works, *e.g. Twenty Thousand Leagues Under the Sea* (1886) by Jules Verne and *The Invisible Man* (1897) by H.G. Wells. As a result, numerous movie interpretations of their work will be discussed throughout this book.

1.3   THE BIRTH OF CELLULOID

The subsequent progression from nitrocellulose to the translucent celluloid needed for motion picture film is full of contention and several claims of patent infringement. The addition of camphor (a waxy solid) and alcohol to nitrocellulose creates a durable, flexible, and translucent plastic known as celluloid, or gelatine as it was then called. In 1884 the American entrepreneur George Eastman and camera inventor William H. Walker patented the invention of gelatine film rolls and the system for using them in photography. This consisted of coating the nitrocellulose plastic with an emulsion of light-sensitive silver halide crystals dispersed in gelatine as a colloid.[5]

Then in 1888 Eastman patented and released the first Kodak camera, which could take 100 photos on a roll of his film. However, early versions of the film produced grainy images, so he turned to Dr Samuel A. Lattimore, head of chemistry at the University of Rochester, to investigate whether it could be improved further. Lattimore recommended the services of his undergraduate assistant Henry Reichenbach. Eastman subsequently hired Reichenbach to improve the chemistry of making film rolls for mass production and they patented the process together in 1889.[6]

In the same year that Louis Le Prince disappeared (1890), Thomas Edison and his assistant William Dickson also announced the development of the first motion-picture camera called the Kinetograph as well as a viewing apparatus known as the Kinetoscope. The Kinetoscope was a mechanical device with a peephole viewer which allowed a single person to watch a strip of Eastman's celluloid film as it moved past a fixed light. In the years following Le Prince's death, his son Adolphe took Edison to court to decide who the true inventor of motion pictures was, but Edison eventually won the case.[3]

Edison and Dickson produced hundreds of short films for the Kinetoscope, one of which is entitled *Mr. Edison at Work in his Chemical Laboratory* (1897). This 20 second clip was actually filmed in a studio build by Dickson, with the benches and chemical supplies brought in from the nearby laboratory building.[7] The camera is in a fixed position as we watch Edison in a lab coat mixing a substance in a bowl near a Bunsen burner and tripod. He is surrounded by chemical containers and bottles, with some familiar laboratory apparatus shown, such as retort stands, funnels and beakers set up in front of him. It finishes with Edison pouring a substance through a large funnel containing a filter paper.[‡]

Other patent infringement cases about the birth of moving pictures would also occupy Edison, Eastman, and many others for decades. Reichenbach and other former Eastman employees eventually formed a rival film-manufacturing company in 1891 called the Photo Materials Company, using new chemical techniques to produce higher quality films. They also subsequently got embroiled in patent infringements with Eastman, went out of business and were eventually merged with the Eastman Kodak Company, which was incorporated in 1892.[8]

Meanwhile, in France a well-known entrepreneur who ran a photographic plate factory decided to improve on Edison's Kinetoscope to bring it to wider audiences. Louis Lumière and his brother Auguste patented their Cinematographe in 1895. It was a combination of film camera and projector that could display moving images on a screen for large audiences to enjoy together.

[‡] Some of Edison's Kinetoscope movies are freely available online, including *Mr. Edison at Work in his Chemical Laboratory* (1897). There are three possible names for *Annabelle's Serpentine Dance*, which is usually associated with the manual colouring process; *Annabelle's Skirt Dance* seems to be associated with the Australian release of the Edison Kinetoscope, and *Annabelle's Butterfly Dance* is associated with the New Zealand release.

The inspiration for the new combined device was the mechanism of a sewing machine. The first ever commercial film screening took place in the Grand Café in Paris in March of 1895 where they showcased several short films of workers leaving their photographic plate factory. The Lumière brothers subsequently opened cinemas across France in 1896, hiring cameramen to create films about daily French life to show to a paying audience.[9]

Colour movies also appeared for the first time during this era which was achieved by Edison's lab assistants painting the films frame by frame. The most famous example of this is *Annabelle's Serpentine Dance* (1895).[‡] The short film shows a woman (Annabelle Whitford) dancing in a long, flowing dress, with sticks in the sleeves to help the garment billow and move. As she dances, the dress shifts between different colours. A homage to this would appear later in the biographical drama about the life of Marie Curie in *Radioactive* (2019).[10]

However, it is the Lumière brothers who are credited with developing and patenting the first practical colour photography process in 1907. It was called the Autochrome Plate, or the Autochrome Lumière. Interestingly, Louis Lumiére also spent his later years working on stereoscopic images, which he called 'anaglyphs', better known today as three dimensional (3D) films. One of the first short 3D movies was called *Train Entering a Station* (1935), which he presented to the French Academy of Sciences. The use of a train moving towards the camera was purposely chosen since a similar scene was used in one of their first public cinema screenings in 1895, which is said to have terrified spectators.[10] Along with references to various works by Jules Verne, homage was paid to *Train Entering a Station* (1935) during the Opening Ceremony of the Paris 2024 Olympics using the original camera footage.

However, it is now generally accepted that the first colour motion picture projected in Kinemacolor is *A Visit to the Seaside* (1908). Directed by George Albert Smith in Brighton, it shows several short clips of people enjoying the beach, dancing, walking, and travelling by horse-led trams.[§] The same technique was

---

[§]*A Visit to the Seaside* is widely available on YouTube and is a rare example of a colour silent film. Since the development of sound on films came after colourisation, there was a short period of cinema history where colour films existed without sound (other than music).

later used for one of the first feature length colour movies called
The World, the Flesh, and the Devil (1914), which premiered in
London as part of a Kinemacolor season.

Finally, the addition of sound to films is usually credited to the
French inventor Eugene Augustin Lauste, who worked with
Thomas Edison in the US on the first Kinetoscope (movie projector) in 1888. However, synchronised sound did not make it
into commercial film until much later (1930s) due to numerous
difficulties with the recording equipment, amplification of the
sound and synchronising it with the visual film.[9]

1.4   THE CELLULOID PROBLEM

Unfortunately, many important examples from early cinema
history have been lost to fire. The chemical breakthrough which
led to the creation of flexible celluloid (nitrate) film was also one
of the biggest problems in early cinema. The addition of camphor wax and alcohol did not subdue the spontaneous flammability of the nitrocellulose-based polymer. One of the earliest
uses for celluloid was aimed at reducing reliance on ivory, due to
shortages caused by overhunting. Eventually it became a substitute for ivory-based billiard balls, piano keys, jewellery, dolls,
and kitchen items. Until the invention of acetate film in the
1950s, it was also for over half a century the only feasible way of
recording movies and photography.[11]

Although the development of nitro-celluloid plastics was a
breakthrough for making movies, its combustibility was a well-
known issue. In fact, if two celluloid billiard balls hit each other
with enough force, they would produce a mild explosion that
would result in "every man in the room pulling his gun". However,
heat and open flames were the biggest problem, in particular the
heat from a candle, lamp or projector mechanism that would
build up over time. With a self-ignition point of only 130–150 °C,
a film showing of significant length came with increased dangers
as the heat built up in the projection system.[12]

The initial screenings of short movies by the Lumiére brothers
and others eventually led to full feature-length films by the turn
of the 20th century. One of the worst disasters associated with
the celluloid problem occurred in 1926 in a temporary cinema
that was set up in the loft of a hardware shop in Dromcollogher,

Ireland. During a double film showing of a short animation called *False Alarm* (1923) followed by a feature film called *The Decoy* (1916), a fire broke out in the projector due to a candle being knocked onto one of the exposed film reels.[¶] Tragically, 48 people (including many children) lost their lives in one of the worse fires in Irish history. Condolences for the Dromcollogher Cinema fire came from far and wide, including King George of Great Britain, President W. T. Cosgrave of the new Irish Republic, and other heads of state.[13]

If flammability wasn't bad enough, longevity was also an issue for celluloid film since it deteriorates *via* thermal, chemical, photochemical, and physical means in a process known as 'celluloid rot'. As it ages the waxy camphor molecules separate from the nitrocellulose matrix due to the slow release of the pressure used during the manufacturing process. This causes the nitrocellulose to crystallize as the camphor molecules build up on the surface. Once exposed to the environment, camphor can undergo sublimation at room temperature, leaving only brittle and flammable nitrocellulose remaining. When it catches fire, nitrocellulose burns fast and hot. It is also virtually inextinguishable because it creates oxygen as it burns, and if stored without adequate venting to release pressure (due to nitrous and nitric oxide build up) it can cause substantial explosions. In addition to all this, the fumes produced when it burns are noxious and can sicken anyone in the vicinity.[11]

Even a hot day can cause a fire, which occurred during the New Jersey heatwave of 1937 in a storage vault and again with the US National Archives fire in 1978.[16,17] One of the most famous titles now considered lost is the first screen adaptation of *The Great Gatsby* (1926). In chemistry terms, *The Schemers* (1922) represents one of the most significant losses to early cinema since it is widely considered to be the first depiction of a black chemist onscreen. The plot follows his work with a drug company researching a synthetic substitute for gasoline. The story revolves around criminals who try to steal Paul's chemical formula, but

---

[¶] Eerily the short animation *False Alarm* is about a clown who tries to put out a fire but runs into numerous difficulties. The days leading up to the Dromcollogher fire were particularly dry, meaning local wells had very low water levels. This made it very difficult for fire crew to put out the cinema fire, on top of the fact that the nitrate film produced its own oxygen when burned. *False Alarm* is freely available online.

little else is known about the onscreen chemistry. Similarly, the US serial *The Black Secret* (1919) is also considered lost with only fragments remaining. We know from stills that there was a depiction of a lab in episode 13, which shows a mortar and pestle, a reaction flask and a scientist in a lab coat looking through a microscope. However, nothing more is known about it.

The flammability and reference to the gelatine coating on nitrocellulose film is expertly portrayed in the Oscar winning Italian movie *Cinema Paradiso* (1988). Early in the story, the projectionist, Alfredo, explains to young Toto that if the 'nitrate film' goes on fire, then you go boom and turn into charcoal. He also explains that if he doesn't move the hand-cranked projector quick enough, everything gets hot and might catch fire. Alfredo also explains that only one side of the nitrate film has the 'gelatine' coating, which Toto says tastes good as he licks it. In reality he wouldn't be able to taste gelatine since it is flavourless food ingredient. However, the gelatine film coating also contains silver nitrate mixed with potassium bromide and/or sodium chloride to create the images, so perhaps Toto likes the salty taste. It should be noted here that licking this type of film is not advisable since silver nitrate is an irritant.

Unknown to Alfredo, Toto steals the explicit film clippings that are cut from every film on the orders of the local priest. This collection of short film clips later catch fire in his house, foretelling the larger fire which eventually destroys the titular cinema and burns Alfredo badly, leaving him blind. It then comes down to Toto to keep the cinema going. Later we see him holding up a match to the new acetate film in the 1950s, demonstrating that is it no longer flammable to which Alfredo bemoans "progress always comes too late".

Cinema Paradiso is widely regarded as reviving the ailing Italian film industry at the time of its release. It is a movie of nostalgia and an ode to the film theatre, especially early movies. Long before the phone and the internet connected people, the cinema brought communities together with scenes of foreign places shown in your local town. This was especially true for silent movies since many people commonly had full conversations during a showing since there was nothing to hear from the movie other than some accompanying music. The movies provided an escape for the cinema goer, allowing them to forget the harsh realities of life if only for a short time. *Cinema Paradiso* (1988) captures the communal experience that cinema provides, celebrating the pure escapism and romanticism of film.

1.5   SCIENCE ON SCREEN

The history of modern science is roughly divided into three eras: the first, in the 17th century, known as the Scientific Revolution; the second, in the 19th century, often referred to as the second Scientific Revolution; and the third, in the first quarter of the 20th century, when many previous theories were overturned, *e.g.* the structure of the atom. The birth of movies coincided with the third era alongside substantial scientific discoveries which captured the public's imagination. For example, the first Nobel Prizes were awarded in 1901, the synthesis of ammonia was scaled up in 1908 (Fritz-Haber process), Rutherford discovered the atomic nucleus in 1911, Einstein published the general theory of relativity in 1915, insulin was isolated in 1921, and Fleming discovered penicillin in 1928, along with many other examples.

The late 19th and early 20th century discoveries transformed science, both as a body of knowledge and as a social and political force. For the first time, it was now necessary to revise curricula and textbooks several times in a generation.[14] Scientific discovery was also accelerating with each new discovery, making it impossible for any one person to maintain a complete knowledge of the latest developments. This fuelled further scientific curiosity as well as fear, especially after WW1 in which toxic chlorine and mustard gas were used for the first time. As a result, it should come as no surprise that early movie makers chose stories involving scientific discovery, hoping to capture the public interest and ensure a financial return for their investment. Constructing multiple sets proved costly and financing was limited due to a fear that the money would not be recouped from this new form of entertainment.

The silent film era provided a unique medium to tell stories entirely through imagery while also freeing stories from the confines of languages for the first time. As a result, the lack of sound may have contributed to the early success of movies, since international distribution was relatively simple from a language perspective. Instead, silent-film actors emphasized body language and facial expression so the audience could understand what an actor was portraying. The melodramatic acting style was in many cases a habit transferred from stage experience, but it was also due to the lack of sound. Also, the onscreen text cards could be translated relatively easily for different regions, giving silent movies a worldwide audience.

Early movie makers drew heavily upon popular stories from famous Victorian novelists for inspiration. The French silent short film *The Invisible Thief* (1909) follows a man who buys the novel *The Invisible Man* by 'G. H. Wells' at a bookshop, although it is unclear whether the author's name is a typo or intentional. In the book he finds a recipe for invisibility and opportunity then makes the thief one of the first onscreen depictions of invisibility. In the real world, the *Invisible Man* by H. G. Wells was published in 1897. Unlike books and radio, the concept of invisibility lends itself well to visual mediums, providing early movies with a differentiator. The popularity of the story spawned numerous adaptions in nearly every genre, even making its way into animation as *The Invisible Mouse* (1947), staring the famous cat and mouse duo Tom and Jerry.

Although no laboratory or glassware is shown in *The Invisible Thief* (1909), it represents the beginning of a theme which goes on to dominate early onscreen chemistry. This is commonly referred to as a 'fear of science' in other works, but this description is far too broad. The theme actually appears more akin to 'research is potentially dangerous' or 'fearing the potential of science'. As mentioned previously, science at the time was advancing so quickly that people could not keep up with the announcements of new discoveries. This created a fear of the unknown, with many fearing that scientists would 'go too far'. More specifically, people feared that scientists had not considered all the possibilities and consequences of their work.

This theme is prominent throughout this book and, it must be said, still exists today in relation to recent advances in artificial intelligence (AI). From a chemistry point of view, we have the term 'chemophobia', but this is usually associated with a 'fear of chemicals', rather than a fear of new advancements in chemistry. The term 'research-phobia' could be applied, but that doesn't really take into the account the fact that some of the warnings later prove to be true when real-world industrial accidents occur. So perhaps the term should be 'research-warning', which will be used in this book in conjunction with 'research-phobia'.

A more explicit early example of this theme comes from the well-known story of Dr Victor Frankenstein by Mary Shelley. The first screen adaption of the popular story is *Frankenstein* (1910) by Edison Studios and begins with a young Frankenstein leaving for college. In a brief scene we see him working in his personal lab or workshop with some glassware and other lab items dotted around his bench. It is explained through an onscreen text card that he has discovered the secret of life and death, and that he intends to "create into life the most perfect human being that the world has yet known". The intentions of Frankenstein are therefore noble, with no indication that he is anything other than a curious scientist. However, his scientific discovery results in an unforeseen side effect when "instead of the perfect human being, the evil in Frankenstein's mind creates a monster".

A further example of this theme comes from another famous gothic novella which also explores the concept of good and evil within us called *The Strange Case of Dr Jekyll and Mr Hyde* by Robert Louis Stevenson. The first onscreen example is a silent short called *Dr Jekyll and Mr Hyde* (1912), which is actually the second US movie about the infamous doctor. Unfortunately, the original 1908 version is now considered a lost film. Previous reports note that the 1908 version focussed more on the horror elements of the story, whereas the 1912 version downplays the horror element in favour of the thematic conflict between the good and evil sides of one's personality.[7]

The 1912 version opens with a pamphlet explaining that the "taking of certain drugs can separate man into two beings – one representing EVIL and the other GOOD". It then features one of the first onscreen images of a full laboratory with numerous bottles on shelves in the background as well as mortar and pestles. The titular doctor is seen working at a lab bench, mixing a powdered solid from a beaker with a solution from a dark bottle into a tall Berzelius beaker or 'pouring glass'. After consuming the concoction, he immediately falls ill and changes into the evil Mr Hyde, expressing delight at his success. Mr Hyde then mixes another concoction in the lab as an antidote, returning him to Dr Jekyll again. After this success, like any good scientist, Dr Jekyll writes down some notes in his experiment notebook.

Sometime later, it is explained through a text card that "repeated use of the drugs causes him to change to his evil self against his will". This results in his scientific discovery (Mr Hyde) causing harm to himself and those around him, *i.e.* scientific discovery leads to disaster. Eventually he runs out of the antidote, resulting in Mr Hyde taking over permanently. With the police and others closing in on him, he destroys the lab in frustration and eventually drinks from an unlabelled bottle to end his life.

It is thought that there are over 150 different stage, film and TV adaptions of the Dr Jekyll and Mr Hyde story, with varying amounts of onscreen chemistry. Another short film version was released in 1913 but it doesn't contain much in the way of onscreen chemistry; instead, Jekyll consumes the concoction in what looks like a medical office. The first full feature length version is a silent movie from 1920. It opens in a lab with a description of Jekyll as an "idealist and philanthropist... by profession, a doctor of medicine". The lab is significantly more complex than previous versions, stocked with hundreds of chemical bottles, glassware, and other equipment. The story begins with a 'research-warning' as the doctor (John Barrymore) and a colleague observe microscopic images of what appear to be cells. His colleague exclaims "Damn it, I don't like it, you are tampering with the supernatural" and tells Jekyll to "stick to the positive sciences".

The 1920 movie emphasises Jekyll as a good person through numerous text cards: "Henry Jekyll is the finest man I know – although we differ on every scientific point". In this version of the story, he is bullied into carrying out the research by other characters after receiving comments like "A man cannot destroy the savage in him by denying its impulses. The only way to get rid of temptation is to yield to it" and "What's the matter, afraid of temptation?". It also emphasises that he is a victim of his scientific discovery, which was spurred on though bullying by others. The creation of Mr Hyde only stems from his desire to separate the evil from the good, so the "evil does not harm the soul of the good person", absolving Jekyll of wrongdoings with "a sea of license".

1.6   REAL SCIENCE

However, it wasn't all gothic horror for early onscreen chemistry: there are also some early examples of themes that would transcend many decades. The German production *Irrungen (Mistakes)* (1913) follows Maria (Henny Porten), a laboratory assistant in a factory research lab. Her father, who also works in the factory, is killed on site and she blames the company management for their lax safety practices. There are various scenes depicting Maria in a realistic lab, mixing solutions, and surrounded by lab glassware from this period (Figure 1.4). She and her colleagues are not only the first female scientists on screen, but also the first scientists of any gender shown in a realistic industrial lab.

The lab features a wide array of glassware on the benches, which includes early versions of volumetric flasks, filter funnels, and breakers. A volumetric flask is a pear-shaped lab apparatus with a round flat bottom commonly used for dilutions and preparing standard solutions. It was invented in 1892 by James Dewar who also invented the vacuum flask still used today and sometimes called a Dewar flask. Along with beakers, condensers and round-bottomed flasks, volumetric flasks will later become a

HENNY PORTEN in ihrer Rolle   „Irrungen"

Figure 1.4   A scene from *Irrungen* (*Mistakes*) (1913), one of the first realistic scenes in a chemistry lab shown onscreen. © Lebrecht Music & Arts/Alamy Stock Photo.

main staple of onscreen labs. The theme of lax safety practises in industrial settings and profit-focussed management will also appear several more times in relation to onscreen chemistry, particularly in the second half of this book.

The emerging field of forensics for crime fighting at the end of the 19th century also provides some early examples of onscreen chemistry, including the use of chemicals (drugs) to manipulate consciousness. The first example of this is a serial of 14 short silent movies called *The Exploits of Elaine* (1914), which likely inspired the *Nancy Drew* series of novels which came later. Elaine works with a jack-of-all-trades scientist who is nearly always shown in the lab filled with glassware and chemicals. He also has a wide range of capabilities which includes the creation of makeshift defibrillators and wire taps, giving *MacGyver* (1985 and 2016 reboot) a run for his money.

In one episode Elaine is injected with scopolamine and the perpetrator forces her to write a letter without her consent or knowledge. Scopolamine is a real-world anticholinergic drug used as medication to treat motion sickness. However, it also interferes with the neurotransmitter acetylcholine, preventing the consolidation of short-term memories into the long term. It is this mechanism that induces amnesia, leading to significant memory gaps. While scopolamine doesn't necessarily enable the perpetrator to control the victim's mind like a puppet master, the induced state of confusion and compliance can make the victim more susceptible to suggestion.

Worth a brief mention here also is an early onscreen adaptation of Arthur Conan Doyle's famous private detective, *Sherlock Holmes* (1916), as a four-part silent drama. In terms of onscreen chemistry, it starts with a scene where a sick man is brought to a chemist's shop where the chemist gives him a tonic. Unfortunately, no details about the tonic are provided. Secondly, Holmes's archnemesis Prof. Moriarty attempts to trap Sherlock by preparing "a sealed chamber for Holmes, to be filled with poison gas at just the right moment". It should come as no surprise that the theme of poison gas rises to prominence after WW1 in several onscreen chemistry examples, especially since WW1 is commonly referred to as 'the chemists' war'. Interestingly, the poison gas is also implied to be flammable since one of the characters stops another from lighting their pipe because it will "blow us all up".

Thankfully the famous sleuth escapes the gas chamber and rescues the kidnap victim in the process. This early screen adaptation of Sherlock is significant because it was long thought lost until a copy was discovered in France in 2014 and restored.

1.7   SILENT SCI-FI

One of the first examples of onscreen sci-fi is the French satire *A Trip to the Moon* (1902), likely inspired by the 1865 Jules Verne novel *From the Earth to the Moon* and the H. G. Wells serial *First Men in the Moon* from 1901. The plot is a light-hearted satire criticizing the conservative and overly serious scientific community of the time. It opens with a meeting of an astronomy club who are all dressed as magicians. But once they decide to travel to the moon, they switch their pointy hats and cloaks for top hats and formal dress, mocking the transition from alchemy to chemistry. Also noteworthy here is the Danish movie *The End of the World* (1916), which is one of the first movies about an asteroid hitting the earth, inspired by the passing of Halley's Comet in 1910.

We also have the famous German movie *Metropolis* (1927), directed by Fritz Lang, which is widely recognised as a sci-fi masterpiece. It is set in a dystopian futuristic city in the year 2026, with stunning visuals even by today's standards. The story follows the son of a wealthy industrialist who lives in luxury among the skyscrapers to the detriment of the workers who live underground in harsh conditions. The story centres around a fictional humanoid robot called the *Maschinenmensch* (machine person) produced in a futuristic lab featuring elaborate glassware, tubing, electronics, and bubbling liquids. *Metropolis* (1927) is said to have influenced later work such as Ridley Scott's *Blade Runner* (1982) and *Brazil* (1985), among others. For many years the original 1927 version was thought lost, but a deteriorated original copy was found in Argentina in 2008 and was painstakingly restored for the Berlin International Film Festival in 2010.

Fritz Lang also directed a trilogy of German movies about Doctor Mabuse (from the novels by Norbert Jaques) consisting of *Dr Mabuse, the Gambler* (1922), *The Testament of Dr Mabuse* (1933) and *The Thousand Eyes of Dr Mabuse* (1960). The actor depicting Dr Mabuse, Rudolf Klein-Rogge, also played the inventor and anti-hero (Rotwang) in Metropolis who creates the *Maschinenmensch* with the intention of destroying the dystopian city. Dr Mabuse, on the other hand, is a criminal mastermind and master of disguise who can use 'telepathic hypnosis' as well as 'body transference' to achieve his goals. Interestingly, from a chemistry perspective, the second movie features brief images of notes with references to toxic gas and a proposed attack on *chemische Fabriken* (chemical factories). A chemical formula is also featured in the notes with water (H_2O) and possibly potassium azide (KN_3), but these are difficult to make out.

Another interesting early sci-fi movie from this era is the soviet made *Cosmic Voyage* (1936), also translated as *Space Journey*. Set 10 years in the future (1946), it tells the story of the first crewed exploration of the moon. The filmmakers used stop-motion animation to depict the cosmonauts bouncing across the low-gravity lunar surface. The well-known real-life Russian rocket scientist Konstantin Tsiolkovsky was consulted during production to ensure accuracy, in doing so possibly becoming one of the first scientific film consultants. This makes *Cosmic Voyage* (1936) was one of the earliest films to depict a realistic spaceflight, including weightlessness.

While exploring the surface, one of the cosmonauts asks why she can't see the Earth. The other explains, through an onscreen text card, this is because they landed on the "side of the moon that can never be seen from Earth", and they can't fly to the other side because their oxygen tanks were damaged. Later, a white powder is found on the surface of the moon, which is said to be the "frozen remains of the Moon's atmosphere". They use it to replace their lost oxygen and fly back to Earth. Interestingly, we now know from the US Apollo and Soviet Luna mission samples (collected 40 years later), that the Moon's surface rocks are in fact made up of approximately 45 wt% oxygen. Albeit the oxygen is tightly bound in various minerals as oxide, meaning electrolysis or a thermal method would be needed to extract it.[15] As a result, it may be the first movie to use a chemical element as a plot device, a concept which we will return to near the end of this book.

REFERENCES

1. M. Ward, Sketches with the Microscope, Offaly History, https://www.offalyhistory.com/shop/books/sketches-with-the-microscope.
2. Nicéphore Niépce House, The Invention of Photography, Nicéphore Niépce's House Museum, https://photo-museum.org/.
3. BBC Education, *Louis Le Prince*, BBC Online, https://web.archive.org/web/19991128020048/http:/www.bbc.co.uk/education/local_heroes/biogs/biogleprince.shtml.
4. Encyclopædia Britannica, Christian Friedrich Schönbein. Encyclopædia Britannica, Inc., https://www.britannica.com/biography/Christian-Friedrich-Schonbein.
5. PBS, *George Eastman*, Why Made America?, https://www.pbs.org/wgbh/theymadeamerica/whomade/eastman_hi.html.
6. The New York Times, Eastman Co. Settles Case, *The New York Times*, 1914.
7. M. A. Griep and M. L. Mikasen, *ReAction!: Chemistry in the Movies*, Oxford University Press, New York, 2009.
8. J. Aumont, Lumiere Revisited, *Film Hist.*, 1996, **8**(4), 416–430.
9. P. C. Spehr and P. C. Spehr, Eugene Augtin Lauste: A Biographical Chronology, *Film Hist.*, 1999, **11**(1), 18–38.
10. D. Rossell, A Chronology of Cinema, 1889–1896, *Film Hist.*, 1995, **7**(2), 115–236.
11. C. W. Saunders and L. T. Taylor, A Review of the Synthesis, Chemistry and Analysis of Nitrocellulose, *J. Energ. Mater.*, 1990, **8**(3), 149–203.
12. K. Eschner, Once Upon a Time, Exploding Billiard Balls Were An Everyday Thing, *Smithsonian Magazine*, 2017, https://www.smithsonianmag.com/smart-news/once-upon-time-exploding-billiard-balls-were-everyday-thing-180962751/.
13. L. Irwin, 'The Calamitous Burning' The Dromcollogher Disaster of 1926, *North Munst. Antiq. J.*, 2013, **53**, 241–265.

14. L. Daston, When Science Went Modern, The Hedgehog Review, https://hedgehogreview.com/issues/the-cultural-contradictions-of-modern-science/articles/when-science-went-modern (accessed 2024-01-09).

15. J. Grant, *The Moon's top layer alone has enough oxygen to sustain 8 billion people for 100 000 years*, The Conversation, https://theconversation.com/the-moons-top-layer-alone-has-enough-oxygen-to-sustain-8-billion-people-for-100-000-years-170013.

16. Fox Film storage fire, *Q. Natl. Fire Prot. Assoc.*, 1937, **31**(2), 136–142.

17. J. Daley, *Forty Years Ago, 12.6 Million Feet of History Went Up in Smoke*, Smithsonian Magazine, 7th Dec, 2018.

Silver Screen

2.1 A NEW ART FORM

In the previous season we discussed how the birth of moving pictures provided a new medium for writers and storytellers. Despite what The Buggles claimed in their song *Video Killed the Radio Star* (1979) when they sang "pictures came and broke your heart", the invention of moving pictures actually predated radio. As we saw in the previous season, moving pictures first emerged in the 1880s. But it wasn't until 1895 that Guglielmo Marconi invented the wireless telegraph while experimenting in his parents' attic. Here he used radio waves to transmit Morse code and, in 1909, shared the Nobel Prize for physics with Ferdinand Braun for the development of wireless telegraphy.[1]

Despite these developments, the first transmission of television signals wasn't achieved until the late 1920s. However, the cameras needed for broadcasting and the cathode ray tube televisions needed to receive the signals were far too expensive for widespread adoption until the 1950s/1960s. Radio was significantly cheaper, and the signals could easily be transmitted from across the world into your house.[2] By the 1920s, radio broadcasting stations became widespread across the world leading to the first multi-episode stories like *The Lone Ranger*, *The Shadow*

Onscreen Chemistry: The Portrayal of Chemical Science in Film and TV
By John O'Donoghue
© John O'Donoghue 2025
Published by the Royal Society of Chemistry, www.rsc.org

and the BBC dramas. But, just like reading a book, the listeners needed to imagine a scene and what the characters looked like based on descriptions.

Back in the world of moving pictures, we already saw how the depictions of chemistry on screen in early movies reflected the fascination and fear of scientific advances in the real world. Although chemical laboratories existed since the late 16th century, the creation of the classical laboratory in the 1860s resulted in a new recognition for the profession of chemistry.[3] Considered a relatively new science after breaking away from alchemy, chemistry was therefore viewed as modern and full of possibilities. However, to ensure that it was taken seriously as a profession, it needed to break all ties with the world of magic, alchemy, sorcery and 'mad science'. This resulted in professional scientists becoming conservative and overly serious. Much like today, the source material for movies came from popular novels, so combining these stories with real-world science resulted in a fascinating blend of horror, fantasy, sci-fi and chemistry.

2.2 THE SOUND OF GOOD AND EVIL

Synchronised sound first appeared in the late 1920s and these new movies became colloquially known as the 'talkies'. This added a new element to onscreen chemistry since chemistry was no longer confined to the visuals of glassware and laboratories; it could now be discussed onscreen as well. Kick-starting our discussion about early talkies is *Frankenstein* (1931) which opens with the doctor and his assistant robbing graves for body parts. They accidentally end up with an "abnormal" brain from the university, which later becomes the source of all their problems, *i.e.* a mistake in their scientific pursuit leads to calamity. The professor at the university describes Frankenstein's (Colin Clive) area of expertise as "chemical galvanism" (electrochemistry) and "electrical biology", which are explicit references to the real-world work by Galvani as discussed previously. This is also a reference to the scientists who influenced the novel's author Mary Shelley, which included work by Galvani and public lectures by Michael Faraday.[4]

Also in the movie, the professor explains that their work was at a very advanced stage, to the point that they were becoming "dangerous", thus giving us an example of the theme of 're-search-phobia' mentioned earlier. However, he also states that Frankenstein "has greatly changed... as a result of his work". The professor explains that the bodies they were using at the university were not "perfect enough" for their work and that Frankenstein insisted on "other bodies". Frankenstein left the university and separated himself from friends and family to complete his work, implying that the isolation, long hours, and obsession with the research has driven him to an apparent madness, although it should be noted that Frankenstein is never referred to as mad.

Except for the test-tube rack on the desk of the university professor's office, the use of traditional laboratory equipment onscreen is minimal throughout. Some glassware is briefly used by Frankenstein to mix solutions, but his lab in the abandoned windmill mainly consists of large electrical equipment with various gauges. Frankenstein is also shown wearing surgical scrubs rather than a lab coat, implying that the experiments are medical rather than scientific. Also, breaking with the estab-lished image of a 'mad scientist' which will be discussed later, he does not wear round safety goggles or have crazy white hair (Figure 2.1).

After the experiment proves to be a success, Frankenstein explains his scientific curiosity to his professor as "Have you never wanted to do anything that was dangerous? Where should we be if nobody tried to find out what lies beyond? Have you never wanted to look beyond the clouds and the stars, or to know what causes trees to bud and what changes the darkness into light?". He finishes by referring to the fear of scientific advancement: "But if you talk like that, people call you crazy". This is one of the first examples linking public misunderstanding of scientific advancement to the concept of madness. However, despite his impassioned defence of sci-entific curiosity at all costs, he never considers the ethics of his work or atones for stealing the body parts to create the new creature.

Frankenstein (1931) was later followed by two sequels in rela-tively rapid succession called *Bride of Frankenstein* (1935) and

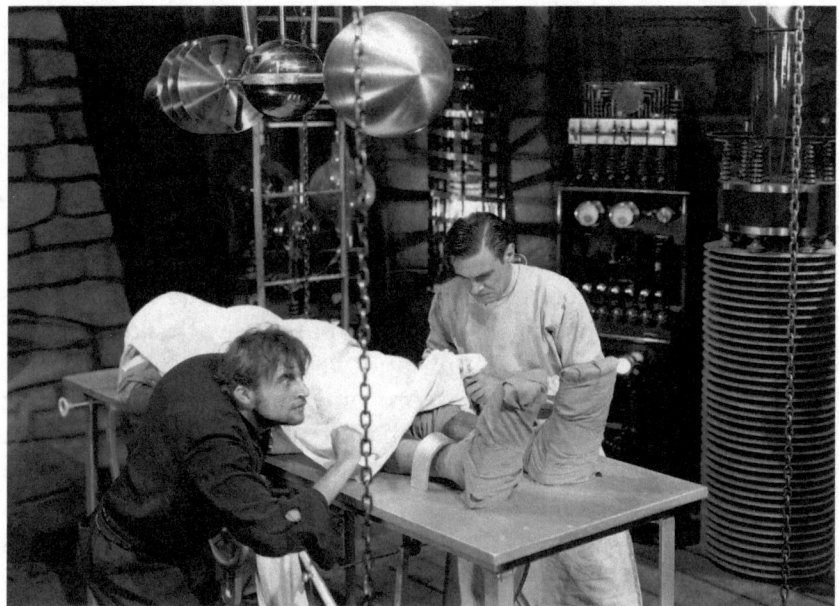

Figure 2.1   The lab from *Frankenstein* (1931) featuring Dr Victor Frankenstein in a surgical gown. Note that he does not have crazy white hair or round safety goggles as are normally associated with 'mad scientists'. © Glasshouse Images/Alamy Stock Photo.

Son of Frankenstein (1939). Again, the depiction of chemistry onscreen is minimal, except for some short scenes involving the mixing of solutions in different types of glassware. There is also a strange conclusion in *Son of Frankenstein* (1939) when the monster is knocked into a conveniently located pit of molten sulfur, situated below the laboratory. We only know that this pit contains sulfur because of an earlier brief mention, but it is not explained how or why this pit exists.

In the same year, we also have *Dr Jekyll and Mr Hyde* (1931), starring Fredric March in the titular role. However, unlike the 1920 silent version, the creation of Hyde is entirely due to Jekyll's scientific curiosity, similar to that seen in *Frankenstein* (1931). The fear of what scientific research might uncover is again at the forefront, with one character exclaiming "There are bounds beyond which one should not go", to which Jekyll replies "I tell you there are no bounds". While pointing at a gas lamp on the street, Jekyll goes on to explain that "if it were not for some man's curiosity, we should not have had it and London would still be lighted by link-boys. And wait, one day, London will glow with incandesence".[†] The other character then replies "I find London quite satisfactory as it is".

The use of a debate between characters to discuss the merits and pitfalls of scientific curiosity in both *Frankenstein* (1931) and *Dr Jekyll and Mr Hyde* (1931) allows the audience to take sides vicariously through the characters.[‡] However, unlike Frankenstein's, the lab depicted in *Dr Jekyll and Mr Hyde* (1931) features an elaborate and active glassware apparatus with liquids bubbling their way through tubing and various flasks. The camera pans along the apparatus from one end of the table to the other where Jekyll is seated wearing a lab coat and smoking a pipe. The camera then switches to a first-person perspective where we see the glassware through Jekyll's eyes, providing a unique perspective for the audience to feel like they are manipulating the glassware themselves. As Jekyll moves his head, the camera follows as he mixes the solutions in test-tubes.

We also discover that Jekyll hasn't slept for days and only "had a cup of tea for breakfast", implying that his scientific obsession is causing concern for his health from those around him; he also misses social engagements due to his work in the lab – all of which establishes the negative side of pursuing scientific research. Interestingly, this version of the story also introduces the use of a mirror in which Jekyll sees himself as Hyde for the first time. This may refer to the real-world chemistry concept of molecular chirality, which occurs when two chemical structures form non-superimposable mirror images of each other. The movie was a critical and commercial success upon its release and was nominated for three Academy Awards, with Fredric March sharing the award for Best Actor.[§]

Wrapping up the theme of Victorian horror monsters and the battle for good and evil, we now turn to the famous 1897 vampire

---

[†] Link-boys were a common sight in London until the introduction of gas lighting in the mid-19th century. A link-boy would carry a flaming torch made from burning pitch and tow to light the way for pedestrians at night, charging a fee of one farthing.

[‡] The use of a fictional debate between characters has been used extensively in the past by scientists to explain their hypotheses, primarily in published text form, *e.g.* Robert Boyle in *The Sceptical Chymist* and by Galileo to explain heliocentrism.

[§] Caution is advised when watching the 1931 version of *Dr Jekyll and Mr Hyde* due to the violence and abuse to women that it depicts at the hands of Mr Hyde.

story by the Irish author Bram Stoker. The first screen adaptation of the story appears as *Dracula* (1931), in the same year as the previous two discussed here. After Dracula's ship reaches England, a newspaper article is shown on screen describing his assistant Renfield as "a raving maniac. His craving to devour ants, flies and other small living things to obtain their blood, puzzles scientists". Later, Prof. Van Helsing is shown in a white lab coat and circular eyeglasses. He pours a clear liquid from a graduated cylinder into a sample of Renfield's blood, which is stored in a test-tube held in place with a retort stand. Immediately the sample in the test-tube turns colourless and then cloudy white as the camera zooms out. Van Helsing claims that this is evidence of a vampire, but it's not clear what chemical reaction has taken place. Other glassware is also shown in the same scene, but they are mostly for visual effect.

The chemistry in *Dracula* (1931) is brief but, unlike other depictions during this era, it serves as a useful tool to find an answer – a concept which gains prominence in the latter half of this book. Van Helsing is not depicted as mad, nor does he contribute in any way to the evil or the horrors. In fact, Van Helsing is described as a "distinguished scientist", saving people with his knowledge of vampires and "strong will". Although, it should be said that references to madness are still prominent throughout, since most of the movie takes place in the sanatorium where Renfield is being held. The image of Van Helsing in a lab coat with round glasses does, however, contribute somewhat to the established image of a 'mad scientist'. A further in-depth discussion about the science of Dracula can be found in *Vampirology: The Science of Horror's Most Famous Fiend* (2021) by Kathryn Harkup.[5]

2.3 THE DANGERS OF SCIENCE

Next, we move onto the first examples of side effects and accidents with the first feature length version of *The Invisible Man* (1933) with synchronised sound. Although this adaptation brought the original H. G. Wells story to the big screen for the first time, it also took some artistic license by introducing the concept of the main character turning insane. It begins with Jack Griffin (Claude Rains) entering a rural pub at night to ask about a room to stay. He is shown wrapped in bandages and wearing dark sunglasses, but his invisibility is only partially revealed to us when he eats some food.

The movie then cuts to a lab with glassware and a scientist in a lab coat using a centrifuge. Referring to a test-tube, Griffin's fiancée tells the scientist to "put that horrid thing down and listen to me". She is worried about Griffin, who has been missing for several days. Dr Kemp explains that Griffin is carrying out one of his own experiments, but "he meddled in things that man should leave alone". This, like other examples discussed here, emphasises the fear and dangers of undertaking scientific research. In an attempt to woo Griffin's fiancée, Dr Kemp claims that "he cares nothing for you, Flora, he never cared about anything, only test-tubes and chemicals". However, this bewilders Flora who explains that Griffin always discussed his experiments with her.

Griffin sets up his glassware, bottles, and books in the room above the pub (Figure 2.2). He then claims that he is attempting to find "a way back", *i.e.* an antidote for his invisibility. Later he pushes away the landlady when she brings him lunch, which

Figure 2.2   *The Invisible Man* (1933) based on the book of the same name by H. G. Wells, featuring a retort flask, test-tubes, and a balance. © BFA/Alamy Stock Photo.

prompts her to tell her husband that he has turned the room into "a chemist's shop". When her husband attempts to remove the lab equipment, Griffin throws him down the stairs. Initially, it seems that Griffin's insanity stems from his frustration at not being able to reverse his invisibility through further scientific research and because of the frequent interruptions to his work. We then see the full extent of his invisibility when a policeman attempts to arrest him, at which point Griffin begins to run amok.

Back in the lab, as colleagues search for clues about his experiments, his boss claims that Griffin never kept secrets. This, along with the earlier statement from his fiancée, implies that his insanity is therefore out of character. As a result, it must be because of the chemicals he was testing by injecting them "under the skin of the arm every day for a month". A list of chemicals is found in his lab with one called "monocaine", described as a "terrible drug.... It draws colour from everything it touches" and when it was tested on animals "it turned the dog dead white, like a marble statue and it also sent it raving mad". Because of the obscurity of the drug and the animal tests, it is concluded that Griffin was unaware of the side effects.

Monocaine is a real-world local anaesthetic first synthesised in 1935 as a successor to cocaine and procaine.[6] Since it was released two years after the release of *The Invisible Man* (1933), it is doubtful that those who synthesised monocaine named it after the chemical in the movie. Also, they probably wouldn't want a new drug associated with side effects like insanity. Interestingly, monocaine is later used as a poison in several murder mysteries like *Matlock* (1986–1995), in the episode The Hucksters (1988), as well as in the TV movie *A Perry Mason Mystery: The Case of the Lethal Lifestyle* (1994) and *The Doctor is Out... Really Out* (2005). But it is difficult to know whether these examples are referencing an overdose of the real-world anaesthetic or the fictitious version of monocaine depicted here.

It is later revealed by Griffin that the motivation behind his research into invisibility was for the benefit of his fiancée, to do something "tremendous", and to gain wealth and fame and honour and to "write my name above the greatest scientists of all time". He exclaims that "I was so pitifully poor. I had nothing to offer you, Flora. I was just a poor struggling chemist". This reference to the financial struggles of chemists will appear again throughout this book as it slowly evolves to match different genres. Numerous adaptions, sequels and remakes of the H. G. Wells story have been made. Modern reincarnations include *Hollow Man* (2000) and *The Invisible Man* (2020), both of which will be discussed later.

A few years later, the debate between the dangers and benefits of scientific progress is also the primary theme of the British movie *Things to Come* (1936). Based on the H. G. Wells novel *The Shape of Things to Come*, it attempts to predict the future from 1940 until 2036. It also explores the dichotomy between the fear of change due to scientific advancement and the need for progression despite this fear. The movie opens with two characters discussing the impact of war on medical research, setting the scene for most of humanity to be killed by war and/or plague by the fictional 1970s.

Human society eventually falls apart, leading to a corrupt warlord (the Chief) taking over the aptly named Everytown. The Chief blames science for everything that has gone wrong in the world and refuses to hear about scientific advancement. When his dictatorship is under threat by those from the outside, he proclaims "Why was all this science ever allowed? Why was it ever let begin? Science, the enemy of everything that is natural in life". This short monologue and the fear of scientific progress demonstrated throughout the movie represent another example of the theme of 'research-phobia' as well as 'research-warning'. That is, scientific research and discovery can cause harm and we should be critical and careful of it.

A turning point in *Things to Come* (1936) occurs when a giant aircraft releases a "gas of peace" over Everytown which acts as an incapacitating agent. This allows the new world order called Wings Over the World to bring peace and scientific development to Everytown. No indication is provided about the chemical makeup of the gas used, but it stands as an interesting visual effect for the real-world interwar period. As mentioned previously, chlorine, phosgene and mustard gas were first used in warfare during WW1, only two decades prior to this movie, and are collectively referred to as "toxic gas" earlier in the story. Gas masks are also mentioned and used several times throughout the movie, providing us with an insight into the influence that chemical warfare had on movie-making during this period. The end of the movie also features a montage of industrialisation and scientific development with giant glassware such as round-bottomed flasks and reflux condensers.

Interestingly, we have another fictional gas in the first screen adaptation of *Buck Rogers* (1939). This 12-part short movie serial opens with the titular captain (Buster Crabbe) flying an airship over Antarctica when they get caught in a terrible storm. Under orders from their superiors, they release the experimental Nirvano Gas, which places them in a suspended animation until they can be rescued. Back in a lab in Westmore Observatory, the gas is explained by a professor through the example of a dog that has been suspended for 3 months. The dog is described as neither "dead nor alive", but his heart is still beating.

However, before Buck and Buddy can be rescued, they get buried in an avalanche for 500 years until the ice has thawed out in the 25th century. This concept was first published in 1910 by H. G. Wells in his novel *The Sleeper Awakes*. In the novel, a man takes some drugs to help him sleep but continues sleeping for 200 years, eventually waking up in a futuristic London where he has become the richest man in the world. References to the novel appear in the US animation series *Futurama* (1999–present), which follows a pizza delivery boy who wakes up 1000 years in the future after stumbling into a cryogenic chamber.

Back to *Buck Rogers* (1939), when Buck and Buddy awake in the 25th century due to the warmer climate, they use a teleporter which functions "by way of radioactivity: it breaks down the atoms of the body into their component parts and, reversing polarity, reassembles them where it desires". No doubt this was a significant influence on the well-known transporter concept which appears later in the *Star Trek* franchise. There is also notable inspiration for the *Star Wars* franchise here, such as the scrolling text at the beginning of every chapter to recap the previous chapter. There is also a brief mention of autonomous spaceships akin to modern drones, but it should be noted that Saturn is falsely depicted as having a solid surface and breathable atmosphere. Also noteworthy is that there is no procedure needed for reanimation: Buck and Buddy are simply thawed out. Nonetheless, the concept of movies accurately predicting the real and fictional future will be revisited again.

In terms of the onscreen chemistry, the benevolent future professor in *Buck Rogers* (1939) is nearly always shown in a lab filled with scientific glassware containing solutions, but he never uses them or talks about chemistry. The overarching theme of the serial, like *Things to Come* (1936), revolves around the 'science age' fighting against the 'traditionalists' who want to "return to the old ways". Buck is on the side of the traditionalists, as is the benevolent professor, interestingly. We will return to Buck Rogers again later and we will also look at the emergence of anaesthesiology in the 19th century, which no doubt influenced the original H. G. Wells novel.

2.4 TIME FOR A CHANGE

It wasn't all monsters and sci-fi in the early talkies: by the mid-1930s we also begin to see a shift away from philosophical debates about scientific research and one-dimensional scientific characters. Instead, we begin to see real labs as well as an exploration of chemists with personal lives outside the lab. The British comedy *The Love Test* (1935) follows Mary (Judy Gunn), who is nominated to be the new head of a chemistry research laboratory. The story revolves around a sexist, rude and dismissive male colleague who attempts to discredit and distract her before she is chosen as the new head, so he can gain the promotion instead. However, another male colleague, Gregg (Louis Hayward), disagrees with his plans, describing her as "a very good chemist, one of the best chemists in the place".

To choose who will woo Mary to distract her from her work, they devise a chemistry experiment akin to drawing straws or lots. They mix some chemicals in a measuring cylinder and order each male colleague to hold a test-tube of the concoction in their right hand: "if it fizzes, you're elected". They work through a few colleagues, declaring "no reaction" as they go, until they set up one test-tube to fizz for Gregg, who begrudgingly accepts. Mary is initially presented as aloof and so engrossed in her work that she doesn't notice what is happening around her. Initially she states that she is only a worker and to achieve this, she has "eliminated sex". However, Gregg's advances and other compliments from her male colleagues encourage her to embrace manicures, flowers, and new clothes.

Interestingly, the research that the team of chemists are working towards is to make cellulose nitrate fireproof, referencing the real-world problem with the film used to make the movie itself. The movie opens in a busy lab with lots of glassware and the scientists are shown in lab coats working on different experiments. The camera pauses on a character pouring a liquid into a measuring cylinder at eye level, which is the correct method to avoid parallax.[¶] Unlike previous depictions of labs, condensers are set up correctly for reflux with two tubes for flowing water in the correct direction, *i.e.* instead of random tubes needlessly joining glassware together in no particular order as seen in *Dr Jekyll and Mr Hyde* (1931).

Most of the movie takes place in or around the lab, but there is also a scene in Gregg's room where he has some glassware and various chemical bottles on his desk. Alcohol is clearly legible on one label and Na_2HPO_4 on the other. This is a real inorganic compound called disodium phosphate and has numerous real-world uses for food. It can be used to adjust pH, prevent coagulation in the preparation of condensed milk, and act as an anti-caking additive in powdered products. Later, the correct formulae for other chemicals are also clearly visible in the lab such as $(NH_4)_2CO_3$ (ammonium carbonate, a leavening agent for food) and $(NH_4)_2Ox$ (ammonium oxalate, an anticoagulant for storing blood outside the body).

Mary is eventually appointed as the head chemist, but her colleagues refuse to work so they can prevent her from getting board approval. Only Gregg continues working on the problem of fireproof film, despite colleagues rigging his experiment to explode. He is eventually successful in making a fireproof coating for celluloid and Mary is appointed as head of the lab. In the real world, the celluloid problem wasn't solved until the 1950s, with acetate films, as mentioned previously in relation to *Cinema Paradiso* (1988). Ironically, *The Love Test* (1935) was long thought lost due to celluloid fire damage, but a copy was thankfully found and restored for the London Film Festival in 1990. It stands today as an important milestone in onscreen chemistry, pre-dating many

---

[¶] Parallax is described as the apparent displacement of an object when viewed from two different points that are not on a straight line with the object. In chemistry terms, this means liquids can be measured incorrectly unless they are measured at eye level, *i.e.* on a straight line.

famous examples by decades. It is also the first example of a female chemist leading a lab, despite the romantic trope that often dominates female characters during this era. It is also the first 'laboratory love' themed movie, which will be discussed again later in relation to *Lessons in Chemistry* (2023). As a result, *The Love Test* (1935) marks the beginning of a golden age of onscreen chemistry, which will be discussed over the course of the next few seasons in this book.

Before we move on though, worth a brief mention here also is *Beauty for the Asking* (1939), which follows a cosmetician who invents a new "astringent" face cream in her kitchen from lanolin and rose water.[‖] As Jean is filling jars with the cream, she exclaims that she feels like an alchemist because the invention "will turn everything into gold". Later, she is convinced to create a new jar because the one she was using "is about as attractive as a can of zinc oxide". Unfortunately, this is the extent of the chemistry in this movie, serving only as a token reference to the role of chemistry in cosmetics. We will return to the theme of cosmetic chemistry later.

2.5   COMEDY CHEMISTRY

Screwball comedy is usually defined as a film subgenre of the romantic comedy genre which became popular during the Great Depression and thrived until the early 1950s. It drew its name from an unorthodox type of pitch in baseball where the ball doesn't travel in a straight line as it approaches the batter. In baseball it describes both an oddball player and "any pitched ball that moves in an unusual or unexpected way". These characteristics were then adapted to describe performers in screwball comedy films, as well as any unusual or unexpected movements. According to Gehring and others, screwball comedy uses nutty behaviour as a prism through which to view a topsy-turvy period in American history.[7]

In the same year as *Things to Come* (1936), the famous Buster Keaton stared in a short screwball comedy called *The Chemist*

[‖] An astringent chemical is a substance that shrinks or contracts body tissues such as skin. Some examples of food items that are astringent include rhubarb and cashews. In cosmetics they are used to tone skin and make it firm by constricting the pores, *i.e.* the opposite of a moisturiser. So the face cream in *Beauty for the Asking* (1939) is a bit of an oxymoron since it is described as both astringent and a moisturiser.

(1936). It features a chemistry researcher in a lab with a variety of glassware, discussing his eccentric inventions with his superior. It opens with the chemist using a Bunsen burner to toast bread as he explains the "Businessman's Breakfast". His invention consists of a powder that can dissolve everything from a traditional breakfast (bacon, sausage, eggs, *etc.*) into a liquid. This allows you to "have breakfast in one gulp", but the eggshells are "difficult to digest". Another invention makes everything three times bigger and stronger, leading to some fish and a bird growing absurdly large. This is most probably referencing the magic pill in *Alice's Adventures in Wonderland* by Lewis Carroll which was first published in 1865 with adaptations to silent film in 1903, 1910 and 1915. These were followed by the first talkies in 1931 and again in 1933.

For the most part, *The Chemist* (1936) is a comedic look at the theme of 'research-warning', poking fun at the numerous examples discussed here earlier. It also pokes fun at the numerous depictions of scientists who don't consider the ethics or consequences of their work. For example, it features several flawed inventions which all have unexpected outcomes. Our chemist also creates a (very inappropriate) chocolate which makes people fall in love with the first person they see, resulting in numerous mishaps. But his greatest invention is a new "noiseless explosive powder" that is activated by water. His excited boss exclaims "This is the greatest invention of all time. This will make blasting a pleasure, now we can have war in peace and comfort." But it also gets noticed by bank robbers, who want to replace nitroglycerin for opening bank safes without drawing attention.

The titular character in *The Chemist* (1936), for the most part, is full of good intentions. But, like the movies discussed earlier, his inventions lead to unexpected consequences, usually with a detrimental effect on the inventor. The movie is also centred around a college, with the characters running in and out of other classes, making this one of the first examples of chemistry in a place of learning. This genre will gain significant prominence later, particularly in the latter half of this book. Eventually the chemist turns the table on the bank robbers, covering them in the explosive powder and marching them down to the police station before it starts to rain. However, just before we think something has finally worked out, he confuses his powders and turns all the college professors small.

2.6   EARLY ANIMATIONS

The first feature film with colour animation was *Snow White and the Seven Dwarfs* (1937), which of course features the famous poison apple. There are several theories behind the use of an apple in this story, which include the forbidden fruit (often depicted as an apple) in the biblical story of Adam and Eve. However, it is mostly likely used in this story because the sweet taste of the apple can hide the taste of the poison. In real life, apple seeds contain amygdalin, which can breakdown into cyanide. However, the amount of cyanide found in apple seeds is generally harmless, unless they are consumed in enormous quantities.

In the movie, the evil queen uses a basement lab in her castle to disguise her image and create the poisoned apple. The lab features several retort flasks (Figure 2.3), tubing, bubbling liquids, and a shelf of books which establishes for the audience

Figure 2.3   An example of a retort condenser flask which features in many early depictions of onscreen chemistry, on display at the School of Chemistry, Trinity College Dublin. Its unique shape allowed a chemist to boil a liquid for distillation or reflux.

what type of lab it is. From left to right the books are labelled *Astrology*, *Black Arts*, *Alchemy*, *Witchcraft*, *Black Magic*, *Disguises*, *Sorcery*, and *Poisons*. She proceeds to uses a retort flask to boil an orange liquid over an oil burner and pours another liquid from a measuring cylinder into a drinking glass. This creates a potion that will disguise her appearance. Lightning, wind and magic spells complete the scene inspired by *Dr Jekyll and Mr Hyde* (1931) as she transforms into the 'witch' or the 'old hag' as she is somethings referred to.

Later, in a cloud of smoke, she dips a yellow apple into a bubbling green brew and when it appears again, the liquid forms an image of a skull on the surface of the glowing red apple. She then cackles "look, on the skin, the symbol of what lies within". This is a reference to the fact that the skull-and-crossbones image was used in real life (and still often used) to symbolise poison. It is also implied that the apple turns red to look more tempting to Snow White. This early example of an onscreen lab exaggerates what we've already seen in relation to *Dr Jekyll and Mr Hyde*, but it also adds magic to real scientific equipment and procedures. The antidote is shown written in the book as "The victim of sleeping death can be revived only by love's first kiss".

Noteworthy here also is *Fantasia* (1940), which features a segment containing potions and magic. A sorcerer called Yen Sid (Disney spelled backwards) is exhausted after practising some spells. He goes for a rest and leaves his magic hat behind. Mickey Mouse, who plays the role of the sorcerer's apprentice, uses the magic hat to command the broom to complete his chores of fetching buckets of water. However, the broom eventually overfills the cauldron and causes a flood. Mickey chops up the boom, but each piece then comes alive and continues the chore. Eventually we see bottles of potions swirling around in the water as the side effect of messing with 'things he didn't understand' floods the entire castle. This is another reference to the concept of not fully considering the consequences of one's actions, linking magic and sorcery to the live action science depictions of the same era.

Next, we have the infamous Acme Corporation, fictional provider of outlandish products that are usually associated with the Wile E. Coyote and Road Runner animated shorts from *Merrie*

Melodies (1931–1997) and *The Looney Tunes* (1930–2014). However, Acme also appears in many other cartoons, commercials, and live-action depictions, usually those made by Warner Bros. Famously, most of their products either fail or backfire at the worst possible times. Although commonly assumed, the name is not an acronym for 'A Company Making Everything'. It actually comes from the Greek word meaning 'the highest' or 'the peak', usually associated with 'the best' or 'the top' of a list.

It started as a joking reference to the Yellow Pages in a phonebook, where people could order anything at the time. Since everything was categorised alphabetically, Acme was always the first option under any category. The name first appeared on screen during the silent era in *Neighbours* (1920) starring Buster Keaton and *Grandma's Boy* (1922). It also appears as the Acme Service Station in the US comedy *Violent Is the Word for Curly* (1938) with The Three Stooges and as the ACME Mining company in the live action western *The Duel at Silver Creek* (1952). Other cartoon depictions include *Cured Duck* (1945) and *Three for Breakfast* (1948), both featuring Donald Duck. Later, the name was also used in relation to several products depicted in *The Pink Panther Show* (1969), one of them being "Pink Pest Control".

In terms of products that Acme supply, the most famous is probably the classic anvil, the natural enemy of many cartoon characters. However, the first Acme chemistry product onscreen is Acme Aspirin, which the Coyote produces from his backpack in anticipation of the inevitable pain he is going to endure as he falls from the sky in *Merrie Melodies – Beep, Beep* (1952). We will return to the Acme corporation again later for more elaborate onscreen chemistry products.

2.7   CHEMISTRY BIOGRAPHIES

The first non-fiction depiction of onscreen chemistry, other than *Mr. Edison at Work in his Chemical Laboratory* (1897), are two biographies about the famous 19th century French chemist Louis Pasteur. The first is a French language biography called *Pasteur* (1935) and the second is an Oscar-winning US biography called *The Story of Louis Pasteur* (1936). Both were produced to celebrate the 40th anniversary of his death in 1895. The French version is based on a 1919 play of the same name and was filmed on location around Paris and Pasteur's hometown of Arbois in eastern France. The film is told as a series of highlights about Pasteur's life and features some brief scenes in a lab with glassware and a microscope. Close up shots of measuring cylinders, Florence flasks, test-tubes, and other containers of liquid are shown on screen. However, it mainly concentrates on his work to convince the medical profession of his germ theory and their pushback.

The US version of the movie opens dramatically with the murder of a doctor for "not washing his hands", resulting in the death of a patient. A pamphlet from Louis Pasteur (Paul Muni) to the doctors and surgeons is shown on screen: "Wash your hands. Boil your Instruments. Microbes cause disease and death to your patients." Immediately the physicians blame Pasteur for the murder and claim that he is a menace to the health profession. Later, when explaining who Pasteur is to the French Emperor, it is also stated that "he isn't even a doctor, sir; he is a mere chemist".

The origin of misinformation is expertly presented onscreen when a surgeon explains to Emperor Louis Napoleon Bonaparte that Pasteur previously heated wine to kill microbes (now called pasteurisation). He incorrectly states that Pasteur plans to "cure blood poisoning in the same manner, namely, by boiling our blood", which horrifies the Emperor. The Empress intervenes in the conversation and explains that the pamphlet said nothing about boiling blood, instead it states that he wants "merely to boil the instruments you surgeons use". But this is rebuked by the surgeon as the practise of witchdoctors.

Also, early in the US movie we are treated to an onscreen 'view through the microscope' of microbes moving, which is a throwback to the microscopic image of cells shown onscreen in *Dr Jekyll and Mr Hyde* (1920). Pasteur exclaims that microbes are killing newborns and over 20 000 women in Paris die annually due to "childbirth fever". In the background we can see Pasteur's lab consisting of various glassware, mortar & pestles, and shelves of Winchester-type bottles. We also get a closeup shot of test-tubes as Pasteur and his colleagues review slides under various microscopes. However, in what might be considered a *faux pas*, Pasteur then looks through his colleague's microscope while still wearing glasses. But, since he removed his glasses for his own microscope earlier, and his colleague doesn't wear glasses, it is likely that Pasteur did not want to adjust his colleague's microscope to compensate for his eyesight.

When his colleagues fail to find the germ that causes "childbirth fever" he tells them to try "again, again and again", emphasising the perseverance needed for great discoveries. This is a theme which we will see again in relation to other biographies. However, Pasteur also misses his supper due to his work, implying that there is some level of obsession, but, unlike *Dr Jekyll and Mr Hyde* (1931), he concedes to his wife and takes a break. Later, his wife also reassures Pasteur that when they have finished their experiments, they will take a holiday, which inspires him to continue. Although he later suffers a stroke, he recovers, and his hard work is recognised with an award. This is an unusual version of the 'research is dangerous' theme, *i.e.* the hard work does affect his health, but he is also rewarded.

The Story of Louis Pasteur (1936) goes to great lengths to demonstrate the conflict between the world of science and that of medicine during the 19th century, presenting scientists as modern, evidenced-based, and hardworking. In contrast, the world of medicine is portrayed as traditional, conservative, and dismissive. The Emperor orders Pasteur to retract the pamphlet, causing him to move home to Arbois where he continues his work through the vaccination of sheep against anthrax (black plague), as depicted in Figure 2.4. This eventually attracts the attention of the chairman of the agricultural board since Pasteur's sheep are the only ones that escaped the plague.

When the minister discovers that it is Pasteur who is responsible, he is again dismissed as "that chemist". Pasteur asks a colleague to explain the process to the chairman, saying "he is a member of the Academy of Medicine, so you will have to use very simple language". The chairman storms off when he discovers that Pasteur is offering the vaccination service free to farmers, implying that the medical profession at the time is solely driven by profit. Pasteur then explains the "life history of the anthrax bacillus in a form that even the layman can understand" to the chairman's assistant through microscope slides. Interestingly, the use of a narrative to explain how the

Figure 2.4   Engraving of Louis Pasteur (1822–1895), French biologist, performing his anthrax vaccination experiment at Pouilly-le-Fort in 1881, which is depicted in the first chemistry biography in the *Story of Louis Pasteur* (1936). Science Photo Library.

germ infects the sheep is now widely accepted as an effective form of science communication.

Pasteur also makes a bold statement to his wife that "the benefits of science are not for scientists, Marie, they're for humanity" and later he also states that "science takes a step, then another, then it stops and reflects, before taking another". There are also further scenes showing correct lab techniques such as filtration, sterilising and measuring, all done by several scientists working together as a team, similar to *The Love Test* (1935). Interestingly, there are also mice, monkeys, dogs, and other animals in cages in the lab for testing, making this one of the first depictions of animal testing on screen. Overall, *The Story of Louis Pasteur* (1936) is the beginning of several excellent biographies featuring onscreen chemistry and real-world chemists during this era.

REFERENCES

1. G. Marconi, *Nobel Prize in Physics 1909*, The Nobel Prize, https://www.nobelprize.org/prizes/physics/1909/marconi/facts/.

2. P. K. Bondyopadhyay, Guglielmo Marconi – The Father of Long Distance Radio Communication – An Engineer's Tribute, *1995 25th European Microwave Conference*, 1995, vol. 2, pp. 879–885.
3. P. J. T. Morris, The History of Chemical Laboratories: A Thematic Approach, *ChemTexts*, 2021, 7(3), 1–18.
4. B. Ruekberg, Another Useful Film Clip: Scientific Methodology of the Frankenstein Monster, *J. Chem. Educ.*, 2021, **98**(12), 4101–4103.
5. K. Harkup, *Vampirology: The Science of Horror's Most Famous Fiend*, Royal Society of Chemistry, 2021.
6. M. L. Tainter and A. H. Throndson, Clinical Results with Monocaine as a Local Anesthetic, *J. Am. Dent. Assoc.*, 1941, **28**(8), 1209–1218.
7. W. D. Gehring, *Romantic Vs. Screwball Comedy: Charting the Difference*, The Scarecrow Press, Plymouth, 2002.

Transition Chemistry

3.1 STRANGE NEW WORLDS

By the end of the 1930s synchronised sound became the stand-ard bearer and colour was starting to appear as well. Two years after Disney's ground-breaking colour animated feature film, *Snow White and the Seven Dwarfs* (1937), a new era began when Dorothy (Judy Garland) opened the door to the technicoloured Land of Oz from her sepia-toned house in *The Wizard of Oz* (1939). Before the era of digital effects, this famous scene was achieved on camera by painting the entire house set in a sepia colour. Judy Garland's stand-in (as seen from behind) also wears sepia-coloured clothes as she opens the door, then, as the camera moves through the door towards Oz, the real Judy Garland with her blue dress steps through the door and onto the yellow brick road. When Dorothy returns to Kansas at the end of the movie, she wakes up in the sepia-toned house again, leaving the colour of Oz behind.

The fascination with other worlds like the Land of Oz can also be seen in the first adaptations of comic strip characters. *Flash Gordan* (1936) is a 13-part serial starting Buster Crabbe as Flash, who also portrayed Buck Rodgers as mentioned earlier. It was promoted with the tagline "Strange World Adventures", which likely inspired the "Strange New Worlds" slogan used in the

Onscreen Chemistry: The Portrayal of Chemical Science in Film and TV
By John O'Donoghue
© John O'Donoghue 2025
Published by the Royal Society of Chemistry, www.rsc.org

Star Trek franchise. The first series was quickly followed by a second larger 15-part serial two years later entitled *Flash Gordon's Trip to Mars* (1938). It is claimed that the 'trip to mars' part was added to the title to capitalise on the panic caused by *The War of the Worlds* (1938) radio drama, narrated by Orson Welles during the Halloween period of the same year. Based on the 1898 H. G. Wells novel of the same name, this ground-breaking and influential story of an alien invasion was not adapted for screen until 1953. In the meantime, the popularity of space heroes and new worlds led to a third 12-part serial entitled *Flash Gordon Conquers the Universe* (1940).

Like *Buck Rogers*, the *Flash Gordan* serials featured a non-descript scientist named Zarkov. However, unlike in *Buck Rodgers*, he is often referred to as 'mad' by the other scientists for his unorthodox ideas. It is interesting that this moniker is used since he is instrumental in saving the Earth which is "doomed to destruction" by collision with the "intensively radioactive" planet Mongo. Zarkov builds the rocketship used by Flash to save the Earth from destruction. There are also some brief scenes showing chemistry glassware filled with bubbling solutions, but there is no real mention of chemistry other than radioactivity. However, of note is the need for a new energy source for the furnaces in the sky city, which use "radium fuel" to generate the "gravity-defying waves". Radium was discovered by Marie and Pierre Curie in 1898 and will be discussed later in relation to *Madame Curie* (1943).

These sci-fi serials often included exaggerated versions of real scientific concepts to move the plot along, much like modern superhero adaptations. In chemistry terms this usually takes the form of a fictional chemical element, which is usually grounded in some reality. In the second *Flash Gordan* serial, the evil emperor Ming the Merciless extracts a fictional element called 'nitron' from the atmosphere to develop a powerful ray gun. In the real world, 'nitron' has been used for several commercial products but it is also associated with the chemical substance $C_{20}H_{16}N_4$ or 1,4-diphenyl-1,2,4-triazol-4-ium-3-yl)phenylazanide. This is a mesoionic compound used for analytical chemistry, specifically for the spectrophotometric detection of nitrates and perchlorates. However, there is no evidence to suggest that it has any ability to function as a ray gun or even a laser for that matter.

Nitrogen lasers, on the other hand, will be discussed later following the invention of lasers in the 1950s.[1]

Unfortunately, strange new worlds required expensive sets and new locations to simulate alien landscapes. By the end of the 1930s they started to disappear from screens as WW2 tightened the purse strings of studios and cinema goers. Budget limitations also delayed the upgrade of equipment for colour, prolonging the use of black and white film until the late 1950s in many cases. However, despite the tighter budgets and significant limitations on travel, this era still produced well-known and critically acclaimed movies like *Citizen Kane* (1941), *Fantasia* (1940), and *Casablanca* (1942), the latter of which was actually filmed on set in Hollywood, despite the exotic location implied in the title.

3.2 TRANSITION CHEMISTRY

It is generally accepted that the 'mad scientist' trope is attributed to the early depictions of science onscreen. However, as we have already seen, the madness nearly always stems from the scientific research rather than the characters themselves, usually because of an unforeseen side effect, *e.g. Dr Jekyll and Mr Hyde, The Invisible Man, etc.* Chemistry in particular was portrayed as being beyond the control of man, a theme which was expertly used for comedic effect in *The Chemist* (1936). The characters are merely victims of their scientific discovery and persistence was often depicted as obsession to the detriment of health. However, it is unclear whether these portrayals were a true reflection of the public perception of chemistry at the time or whether they were due to the source material being used, *i.e.* gothic horror novels.

Thankfully, by the mid-1930s we begin to see a shift away from the philosophical debate about scientific research, gothic horror, and one-dimensional chemist characters. However, when a scientist suggests something that might be perceived as 'fantasy', those scientists are still referred to as 'mad' by other scientific professionals and lay people. Movies like *The Love Test* (1935) and *The Story of Louis Pasteur* (1936) set a new tone for onscreen chemistry in terms of fiction and non-fiction. They accurately depicted lab environments and the chemist was depicted as having a life outside of the lab. In the case of *The Love Test*

(1935), it also depicts accurate chemistry skills when one of the characters accounts for parallax. The scientific conversations are also realistic, even though the chemistry described in the movie was entirely fictional at the time.

A year after *The Love Test* (1935), we briefly return to some fictional forensics in *Charlie Chan at the Race Track* (1936). This was the 16th movie in the *Charlie Chan* series of mysteries which revolved around the fictional Hawaiian detective who is later referenced in the TV series *Hawaii Five-O* (1968–1980). While investigating the death of a racehorse owner who was kicked by his stallion, Chan discovers an international gambling ring. Relevant to onscreen chemistry is a brief scene in a photo-development lab where the technician explains that there is a "photoelectric cell at every quarter mile marker" to time the horse. There is also a "highspeed movie camera" set up to take a photo at the finish line. Throughout this scene there are several chemical containers and funnels shown on the shelves in the background, presumably used for developing the photos. The detailed description of how the cameras work is a possible reference to the work of Eadweard Muybrudge in relation to animal locomotion from 1878, as mentioned earlier.

Later, in his hotel room, the famous detective has set up some glassware, a tripod, and a burner. He then conducts a chemical test of the foam he collected from the mouth of one of the horses. The solution in the test-tube turns cloudy and starts bubbling, during which he claims "bubbling proves the foam contains a very powerful stimulant of the heart". This may be the first example of performance-enhancing drugs mentioned onscreen, but nothing more is explained about the stimulant or the chemical test. These concepts are later modernised and referenced in the James Bond feature film *A View to a Kill* (1985), which will be discussed later.

3.3 MAGIC CHEMISTRY

We now return to non-fiction onscreen chemistry with four excellent biographies, which follows on from our previous discussion about *The Story of Louis Pasteur* (1936). The first of which is the Oscar-nominated US movie *Dr. Ehrlich's Magic Bullet* (1940), about the Nobel Prize-winning German doctor Paul Ehrlich.

Although set during the German Empire at the beginning of the 20th century, it contains numerous references to the Nazi regime which had come to power during the movie's production. For example, Ehrlich declares "what has race got to do with science?" when asked about the Japanese scientist Sahachiro Hata in his lab. The movie was banned in Germany by the Nazi government at the time who were opposed to the movie's tribute to a Jewish chemist. The opening credits contain a dedication to "Dr. Paul Ehrlich whose dream it was to create out of chemicals 'Magic Bullets' with which to fight the scourges of mankind".

The movie opens with Ehrlich (Edward G. Robinson) moonlighting in the hospital lab (while enjoying a cigar) attempting to find a way to stain bacteria so he can observe them under a microscope. Unlike previous depictions of lab research, Ehrlich does not turn mad, and his perseverance eventually pays off for the good of him and others. However, worthy of note is a scene where his colleagues state that he seldom leaves his home lab, implying that he has developed an unhealthy obsession with his research. This obsession with his work eventually results in him contracting tuberculosis (TB) while developing a stain for the bacteria. But, unlike fictional depictions, this setback is only temporary since he recovers and discovers immunization in the process. (Figure 3.1)

There are also numerous references and depictions of other real-world scientists such as Louis Pasteur and Robert Koch, in addition to Sahachiro Hata as mentioned earlier. The movie also includes accurate explanations about aniline dyes (methylene blue) for staining cells and Ehrlich explains chemical bonding as "the attraction of certain atoms for certain other atoms causes them to unite to form compounds". Numerous images of laboratories, apparatus, controlled trials, and references to real-world chemistry are depicted throughout. In fact, *Dr. Ehrlich's Magic Bullet* (1940) sets a new bar for onscreen chemistry with more references than any other onscreen production up to this point. There is also a scene which many researchers even today will find amusing where Robert Koch declares "what do the members of the budget committee know about the requirements of science".

Overall, *Dr. Ehrlich's Magic Bullet* (1940) stands as a groundbreaking departure from earlier depictions of chemistry research

Figure 3.1 The biography *Dr. Ehrlich's Magic Bullet* (1940) was banned in Germany upon its release, but the scientist Paul Ehrlich was later honoured with a stamp in 2004 along with his colleague Emil von Behring. © Olga Popova/Shutterstock.

and even includes a discussion about the potential side effects of chemotherapy. Ehrlich faces numerous barriers throughout his journey, in particular from his colleague von Behring who exclaims "the idea of shooting chemicals into the veins of human beings fills me with horror". Ehrlich sketches out the concept of arsenic-containing molecules shaped like a key that will fit into a microbe without affecting the human body. Eventually, he discovers a chemical cure for syphilis but warns about the dangers of potential side effects and demands extensive trials. It becomes known as compound 606 or salvarsan (arsphenamine), now recognised as the first modern antimicrobial agent.

Towards the end of the movie, all the patients with syphilis ask to be involved in the human trials, despite the warnings, implying that the need for a cure outweighed the potential dangers. However, salvarsan is later linked to some deaths in patients with underlying issues, resulting in a legal trial. At the trial, it is explained that in rare occurrences, when the compound is broken down, the arsenic might be released into its elemental form. The trial concludes when von Behring states that side effects should be acceptable in the interest of eradicating syphilis, vindicating Ehrlich's work. In the real world, Ehrlich's lab developed neosalvarsan in 1912 to replace salvarsan, which was easier to prepare and administer. It found widespread use until penicillin came along in the 1940s.

Interestingly, in the same year as *Dr. Ehrlich's Magic Bullet* (1940), we also have a biography about the life of Thomas Edison called *Edison, the Man* (1940), staring Spencer Tracy as Edison. It centres around the invention of the light bulb, phonograph, and kinescope, as mentioned previously. We are treated to some chemistry, lab glassware, and lab skills when Edison and his co-workers are testing materials to find an appropriate filament for their new light bulb. They state that they have tested dozens of metals like tin, aluminium, copper and iridium. Following this conversation, a montage of lab work is presented showing a variety of scientists using glassware coupled with an image of chemical elements being crossed off in a notebook which includes sulfur, manganese, lead, mercury, antimony, *etc.* Eventually, Edison realises that he needs to remove oxygen from the light bulb which will allow his carbon filament to glow without burning out.

3.4 THE CHEMISTRY OF TOOTHACHES

Another chemistry-themed biography from this era in the same vein as *Dr. Ehrlich's Magic Bullet* (1940) is *The Great Moment* (1944). It tells the story of the 19th century dentist W. T. G. Morton (Joel McCrea), who discovers the world's first anaesthetic as 'sulfuric ether', now called diethyl ether. The movie opens with a text card explaining "before ether there was nothing. The patient was strapped down... that is all". The story begins with a man entering a pawn shop to purchase Morton's medal so he can return it to Morton's widowed wife. It is then her reminiscing and flashbacks which lead us through the story.

The flashback begins with Morton receiving a letter from the US congress who voted to give him $100 000 for his patent. He travels to Washington D.C. but President Franklin Pierce refuses to sign the bill until they sue the Navy surgeon who ignored the patent and claimed to have invented the anaesthetic himself. However, public opinion turns against Morton with newspapers labelling him as greedy, which eventual results in the American Medical Association (AMA) disowning him. After the court rules that the discovery is unpatentable, Morton enters a glassware supply shop and smashes the ether inhalers they are selling without paying him royalties. The result is the ruination of his dental practise, and he dies a broken man.

The flashbacks then continue further back in time to explore the origins of his claims. Morton visits a chemistry professor named Charles T. Jackson (Julius Tannen) at a bar who tells him that the only way to desensitize a nerve is to freeze it *via* evaporation. He suggests using something with a low boiling point like "ethyl chloride drops". When Morton asks him what ethyl chloride is, the professor gives him the correct chemical formula as C_2H_5Cl. Morton then heads to a pharmacy, but the pharmacist asks if he wants chloric or sulfuric ether, giving the correct formula for both (sulfuric is $C_2H_5OC_2H_5$). An intrigued Morton takes home a bottle of both types to run some experiments. Note, the reason diethyl ether is called 'sulfuric ether' in this movie, despite the lack of sulfur in the chemical formula, is because sulfuric acid is used in its creation by dehydrating ethanol.

Back home, while Morton reads a medical book, the sulfuric ether boils near the fireplace. A quote from the book is provided on screen correctly stating that it boils at 35 °C. However, the other bottle containing ethyl chloride has a boiling point of 12 °C, which means it should have evaporated first in this scene. As the sulfuric ether evaporates, the vapours cause Morton to pass out resulting in his wife picking him off the floor thinking he is drunk. Later, Morton is visited by another dentist, Horace Wells, who claims the key to painless tooth extraction is nitrous oxide, better known as laughing gas. But when they take their idea to Jackson, the chemist at Harvard Medical School, he tells them that "Henry Hill Hickman went through all that, Priestly found laughing gas and Humphrey Davy tried all that stuff 50 years ago. Faraday experimented with every type of inhalant he could get his hands on".[†]

He warns them to give up before they kill somebody and tells them to go back to their "tooth yanking and leave science to the scientists". This stands as another prominent example of the conflict between scientists and medical practitioners during the

---

[†] Henry Hill Hickman was an English physician and promoter of anaesthesia. He conducted numerous experiments on animals using carbon dioxide and other gases. Joseph Priestly, Humphrey Davy and Michael Faraday were all famous 18th and 19th century chemists.

19th century, like that seen previously in *The Story of Louis Pasteur* (1935). Despite the warnings, they go on to perform an experiment with a willing participant in front of a live audience, calling it the "Wells' method of painless extraction". They collect nitrous oxide in a balloon attached to a bell jar placed in a water bath. The gas is produced in a boiling flask over a burner, which is presumably filled with ammonium nitrate, but this is not mentioned on screen. Jackson warns the participant not to inhale too much of "that stuff". While inhaling, the participant starts laughing uncontrollably and eventually wakes up during the extraction, rendering the experiment a complete failure. Later Morton is reading through his medical textbooks and the words "Nitrous, Oxide" appears on screen as well as a paragraph referring to Faraday's monograph from 1818: "By the incautious breathing of ether vapour a man was thrown into a lethargic condition which, with a few interruptions, lasted for thirty hours".

This leads Morton to chase his dog around the house with ether. Failing this, he experiments on himself by inhaling ether and impaling his hand on a spike for notes on his desk. Later, he uses a volumetric flask with glass tubes to allow patients to inhale the ether before tooth extractions. However, after a patient jumps out of the chair and "turns mad", Morton visits Jackson in his lab again to investigate the problem. After smelling it, Jackson explains that the ether he got from the new supplier was not "highly rectified" and was instead "cleaning fluid". In chemistry terms, 'rectified' means purified or refined by repeated or continuous distillation. The reference to cleaning fluid is also interesting, since impure diethyl ether can contain methanol, water, acetic aldehyde and peroxides, all of which can be associated with the smell of cleaning agents.

Although *The Great Moment* (1940) is a welcome departure from the 'mad scientist' depictions of the early 1930s, it depicts the chemist (Jackson) as drunk, egocentric, and greedy. Once Morton perfects ether as an anaesthetic, Jackson comes looking for a greater share of the discovery. In response, Morton claims that he is mad when he states "the problem with you, Professor, is that you are cracked in the head. You did not discover ether narcosis, I did". Towards the end of the movie, as Morton is attempting to convince a surgeon to use ether, the surgeon correctly asks for "ammonium carbonate" as smelling salts. The movie also associates scientific perseverance with success, like that seen in the other biographies from this era. Morton states to his wife "it can't have been much fun having a husband who always reeked of chemicals, works all night, never comes home, forgets dinner parties and birthdays". However, although he was successful in his discovery, he was ultimately unsuccessful with his patent application, leaving the couple destitute.

3.5 FICTIONAL SCIENCE *VS.* MEDICINE

So far, we have seen numerous examples of the real-world conflict between the scientific and medical communities. Scientists have referred to the medical community as conservative and greedy, while the medical community have called scientists 'mad'. Up to now we saw how the two communities were usually merged into one for fictional stories like that seen in *Frankenstein* and *Dr Jekyll and Mr Hyde*. However, by the mid-1930s, fictional stories also begin to embrace the conflict between science and medicine. The independent US horror *Maniac* (1934) is a loose adaption of the 1843 Edgar Allan Poe short story *The Black Cat*. It follows Dr Meirschultz and his assistant who are attempting to bring the dead back to life. Note, this is not to be confused with Victor Frankenstein who creates a new living being using parts from different bodies. The prologue in *Maniac* (1934) provides an unsubstantiated statistic that "The Chicago Crime Commission made a survey of 40 000 convicted criminals and found them all suffering from some mental disease".

The movie opens in a lab filled with chemical containers and scientific glassware which includes condensers, boiling flasks, syringes, and test-tubes. The scientist is depicted as manic, with white hair, round glasses, and dressed in a surgical gown. When he asks his assistant to retrieve a body from the morgue to test his hypothesis, the assistant refuses and objects to the morality of the research. They are eventually successful, but the scientist claims their success was too easy since the body was "fresh". He wants a bigger challenge so he asks the assistant to shoot himself so he can test the experiment again. Instead, the assistant shoots the scientist. This is followed by a text card explaining that dementia is "the most important of the psychoses, both because it constitutes the highest percentage of mental diseases and because recovery is so extremely rare".

The assistant then impersonates the scientist so he can hide the murder from others. In doing so, he changes his hair to match the scientist and dons some glasses. He also takes on his eccentricities by laughing loudly and shouting around the doctor's office. What initially starts as an impersonation eventually turns into paranoia as he switches between both personalities to interact with different people. *Maniac* (1934) is one of the earliest depictions of what is now considered to be the standard image of a 'mad scientist', which is emphasised by the numerous links with real mental illnesses through onscreen text cards.

Similarly, a few years later we also have *The Man They Could Not Hang* (1939), which follows a famous scientist who is experimenting with bringing the dead back to life. He claims that a mechanical heart would allow doctors to carry out procedures on dead bodies that would be impossible on living bodies. However, similar to fictional scientists like Dr Jekyll and real scientists like Paul Ehrlich, the scientist (Boris Karloff) claims that he needs to carry out this work in his home lab away from those who would judge him. Several examples of empty glassware can be seen in the lab such as beakers, measuring cylinders, various forms of flasks, and some equipment. Two tanks of gas are also shown, one labelled "oxygen" and the other labelled "carbon dioxide".

While discussing the procedure with his assistant, the doctor compares the tubing in his lab to the arteries connected to the heart. The artificial heart is connected to an elaborate setup of glassware and tubes, all of which are connected to a pump. The idea is that they can use this device to keep the blood pumping after the volunteer dies. Later, the device is shown in operation with a transparent colourless liquid bubbling through the tubing while the artificial heart ticks back and forth like a clock. When asked how he killed the volunteer, he states "I made use of certain gases that end life without poisoning the tissue".

However, before he successfully completes the procedure, he is arrested when the police are alerted by the volunteer's fiancée. During the proceeding court case, the police surgeon represents the professional medical community and the scientist represents those who are trying to push against the establishment to save people with the artificial heart. When the police surgeon is asked why he stopped the scientist from reviving the volunteer, he says "because the idea was fantastic", implying that it was fantasy or unrealistic. The surgeon is also asked about "dead people being revived by drugs", which he claims is impossible. Later, the scientist who is hung for his crime is revived by his assistant using the artificial heart mechanism and through major surgery on his neck. However, angry at how he was treated and frustrated at the misunderstanding of his work, he decides to take his revenge on the jurors and the judge. Therefore, the apparent 'madness' here is a result of frustration and unfair treatment. It is noteworthy that *The Man They Could Not Hang* (1939) was released a year after the first artificial heart was implanted in a dog in the real world.[2]

Next, we return to *Dr Jekyll and Mr Hyde* (1941), this time staring Spencer Tracy (*Edison, the Man*) and Ingrid Bergman (*Casablanca*). Once again Dr Jekyll is depicted as a good and decent man, helping a patient suffering from shock after an explosion. Jekyll requests permission from his superiors to "cure" the man through chemical means, but his request is denied by the hospital because he has only tested his experimental cure on animals to date. The hospital claims that his "chemicals may be deadly" for humans. He responds that "he is no witch-doctor" and that sometimes "we have to gamble", echoing similar statements made in *Dr. Ehrlich's Magic Bullet* (1940). This viewpoint is expanded further when Jekyll implies that the medical council are unaccepting of "advanced theories" and that they only accept ideas that are good for their finances. This also echoes similar implications made about the medical profession in *The Story of Louis Pasteur* (1936).

Jekyll then works around the clock to prove his theory and develop his chemical cure. Like previous depictions, we are shown an elaborate home lab filled with glassware. However, of note is that it also includes animal test subjects for the first time, in the form of rabbits. However, the lab only appears in a few scenes, significantly less than the 1931 version of the story, despite the similarities in the layout of the lab. Like earlier versions of the story, he forgoes lunch, turns up late to dinner, and misses a concert due to his "research work in separating the facets of the brain". In a montage of his lab work which eventually leads to his success, we get a brief look at his notebook which shows HCl (hydrochloric acid) and $NaHCO_2$ (incorrect formula for sodium bicarbonate, it should have a subscript 3).

Later, there is an interesting dinner discussion revolving around the man suffering from shock mentioned earlier. The patient is described as "quite mad", but Jekyll is unconvinced of this assessment. He explains that only one side of him is showing, his evil side, and he wishes to revert him to his good side using his chemical cure. However, before Jekyll can test his cure the man passes away, presumably from suicide, although this is not made clear. This forces Jekyll to test the cure on himself. The chemical concoction is shown bubbling vigorously without any apparent heat source, implying that the movie makers are using dry ice CO_2 for visual effect. Eventually, as before, Jekyll changes into Hyde spontaneously with terrible results. In the latter stages of the movie, Hyde desperately convinces a colleague to fetch the chemicals marked A-M-S and Z from his lab so he can change back to Dr Jekyll. The story once again finishes in the lab with Jekyll/Hyde changing spontaneously and destroying all the glassware as he tries to escape from the police.

In every version that we have seen so far, Dr Jekyll is referred to as a medical doctor and never a chemist, despite the use of onscreen labs and chemical cures. *The Mad Ghoul* (1943), on the other hand, features a university chemistry professor named Dr Morris (George Zucco) who is teaching in a medical department similar to Jackson in *Dr. Ehrlich's Magic Bullet* (1940). He explains that he wishes to recreate an ancient Mayan poison gas, for "research purposes". The movie opens in a classroom with a workbench filled with laboratory glassware. He invites a student called Ted to his research lab in the basement of his house, which includes several monkey test subjects. This reiterates the common theme of animal test subjects featuring in onscreen labs during this era. Dr Morris describes the home lab as his "private universe", implying that he is free to do as he wishes away from university oversight. During the tour of his "private universe", he holds up a stoppered flask filled with a white powder and explains that "these crystals and the gas they give off are colourless and odourless; therefore, more deadly", warning the student to wear a mask when working with it.

The development of an antidote to the poison gas involves the surgical removal of a heart from a donor, implying the need for a medical student. While testing the antidote on a poisoned monkey, Ted exclaims "I can't help feeling a sense of evil in all this" to which Dr Morris responds "moral concepts – as a scientist there is no good or evil, only true or false. I work with one, discard the other". This is a subtle reference to the concepts explored in the now well-trodden *Dr Jekyll and Mr Hyde* story. Later, we learn his true motives as he attempts to seduce Ted's girlfriend, who is described as a famous singer. He conceals the Mayan poison in his desk so Ted inhales it accidentally, turning him into a type of zombie that he can control and manipulate. He then convinces the poisoned Ted that his girlfriend doesn't love him anymore and that she actually loves Dr Morris.

Later, feeling regret, Dr Morris and Ted go to the cemetery to obtain a heart to cure Ted, but it transpires that the antidote is only temporary. Eventually things get out of hand as they look for more hearts from cadavers to develop further antidotes. Overall, Dr Morris is easily the most corrupt and unscrupulous chemist depicted on screen, representing a significant departure from other depictions to date. However, at one point Dr Morris calls himself a physician, so perhaps he also has medical training, but this doesn't explain why he needed a medical student to extract the hearts. Overall, *The Mad Ghoul* (1943) borrows themes from *Frankenstein* (grave robbing and creating a monster) and *Dr Jekyll and Mr Hyde* (good and evil, split personality, and relapsing to an alternate state).

Next, we return to *The Invisible Man* (1933) franchise with two sequels called *The Invisible Man Returns* (1940) and *The Invisible Woman* (1940), both released in the same year. Interestingly, both stories move away from self-experimentation and instead the scientist makes someone else invisible. *The Invisible Man Returns* (1940) tells the story of a coal miner (Radcliffe) who is saved from execution through invisibility by the brother of the protagonist from *The Invisible Man* (1933). Numerous scenes take place in a lab with glassware, a weighing balance, bubbling solutions, and containers.

Interestingly, the invisibility drug from the first movie, monocaine, is now replaced with duocaine, described as an extract of an East Indian herb. Duocaine is a real-world local anaesthetic, but it was not discovered until the late 1950s, long after this movie. The jack-of-all-trades scientist works hard in the lab to find an antidote, using rabbits and guinea pigs as test subjects. However, before an antidote can be found, madness begins to set in again, just like for his predecessor. Radcliffe takes revenge for his false imprisonment and clears his name but ends up significantly weakened after the ordeal. Eventually, a blood transfusion turns Radcliffe visible again, giving us a happy ending this time.

The Invisible Women (1940) on the other hand is a screwball comedy. But, instead of a drug, the atypical 'mad scientist' Professor Gibbs uses a machine in conjunction with an injection to make a clothing model invisible so she can get revenge on her boss. The chemistry here is brief, but there are complaints about the cost of chemicals at the beginning and some glassware can be seen on occasion, especially near the end when one of the thugs gets trapped inside an enormous bell jar. There are also some close-up shots of chemical containers, with the words "rubbing alcohol" clearly visible on one of the labels. The professor's name is also a reference to the real-world American scientist Josiah Willard Gibbs, from which the chemistry concept 'Gibbs free energy' is named after. Interestingly, one of the other inventions by the professor is an autonomous car which can park itself in the garage, a prediction that was well ahead of its time.

Also during this era we get a combination of monsters and mad science in the comedy horror *House of Frankenstein* (1944). The protagonist is a scientist who escapes from prison and recruits Dracula, Frankenstein's monster and the Wolf Man to get revenge on his behalf. The walls of his prison cell are covered in mostly correct chemical structures such as benzene rings, methane (CH_4), water (H_2O), ethanol (CH_3CH_2OH), hydrogen bromide (HBr) and various carboxylic acids, among many others. Later, as he rebuilds his secluded lab to resurrect Frankenstein's monster, we are treated to a lab montage featuring some lab equipment. Later, there is also a brief scene of the scientist working at a bench with some lab glassware containing the harvested organs that he needs.

Finally for this section, we stick with the comedy genre with some brief references to chemistry in *Arsenic and Old Lace* (1944), staring Cary Grant as Mortimer. This is a macabre black comedy based on the Broadway play of the same name and follows two wealthy aunts who poison lonely old bachelors to end "their suffering" and "give them peace". However, despite the title, the chemistry references don't go beyond a single scene where the aunts explain that they obtained the poisons from an old laboratory that once belonged to Mortimer's grandfather. They also explain that the poison consists of "a gallon of elderberry wine, one teaspoon of arsenic, half a teaspoon full of strychnine and then just a pinch of cyanide", to which Mortimer responds "hmm, should have quite a kick".

Although never shown, the old lab in *Arsenic and Old Lace* (1944) is implied to be in the house. In most of the examples discussed in this section, home labs are usually associated with a desire to carry out unethical work away from prying eyes, as seen in all versions of the *Dr Jekyll and Mr Hyde* story as well as in *The Mad Ghoul* (1943) and others. But as noted earlier, they may also be used for developing photos and can allow a scientist to work from home as seen in *Dr. Ehrlich's Magic Bullet* (1940), which results in a positive discovery. However, by the end of the 1940s the home chemistry lab was disappearing from screens in line with the professionalisation of chemistry into universities and industry labs.

3.6   MADAME CURIE

Less than a decade after her death, Madame Marie Skłodowska-Curie was honoured onscreen with the first adaptation of the biographical novel of the same name, written by her youngest daughter Ève Curie. *Madame Curie* (1943) stars Greer Garson in the titular role and begins with Marie fainting in a lecture theatre due to overwork and failure to eat while studying for two master's degrees in physics and mathematics. However, the professor takes Marie under his wing, emphasising that she needs other people in her life like friends and a life partner. He tells her that he also loves physics and mathematics, but he still has a wife, children, and grandchildren. This represents a departure from other movies discussed here, *i.e.* this is the first time that a scientist is told that they need to have a life outside of the lab. He invites her to have dinner with his wife and introduces her to Pierre Curie in the "school of physics and chemistry". Later, Pierre accepts a new assistant into his lab as a favour to the professor without realising that the new assistant is Marie.

The next scene takes place in the lab with a discussion between Pierre and his male assistant, David, about women in scientific research, claiming that they are a "distraction" and the "natural enemy of science". Pierre also claims that "women and science are incompatible" and "women of genius are rare". Upon realising that the new lab assistant is Marie, Pierre then proceeds to fumble over himself to make sure Marie is comfortable and has everything she needs. All the while David is making a lot of noise around the lab, distracting Pierre and Marie. Later, Pierre is in deep thought after a conversation with Marie, which results in a near miss with a horse and cart, foreshadowing his untimely death later in the movie. Not much is mentioned about the work in the lab until Pierre presents his book to Marie entitled *Symmetry and Physical Phenomena: Symmetry of an Electric Field and a Magnetic Field*.

Later, there is a brief depiction of the real-world scientist Henri Becquerel who demonstrates his discovery of radioactivity as energy emanating from rocks. Initially, he incorrectly states that the rocks absorb energy from the sun and then re-emit the sunlight later. However, his experiments are unsuccessful except for one mineral rock, which he calls pitchblende. Marie picks up the rock and exclaims "it is as if there were a piece of the sun locked up in here", which emphasises the theme of 'stars' that are referenced and shown throughout the movie. These references link their discoveries with that of nuclear fusion, the energy source which produces the energy in stars. The conversation about work/life balance also appears again when Pierre tries to convince Marie to stay in Paris, "I do not understand how anyone with a scientific mind can entertain the thought of abandoning science", to which she responds "I know, but there are other things that are important too".

The second half of the movie moves from physics to chemistry as Marie and Pierre begin their work on the isolation of radium (Figure 3.2). The lab expands with more glassware and Pierre acquires an "electrometer". This is a real instrument for measuring the electric current pulses which are induced by radiation between two electrodes in a gas-filled chamber. Uranium and thorium are identified as the elements that give off the rays, but the radiation from the pitchblende does not match the amount of uranium and thorium present. Pierre asks

Figure 3.2   Real images of Marie and Pierre Curie working in their lab, such as this one, helped to ensure *Madame Curie* (1943) was accurate in its depiction of the famous chemists. © Wellcome Images/ Science Photo Library.

to see the chemical analysis of the pitchblende, which Marie has on a chalkboard.

Uranium oxide normally has the formula UO_2, but the formula provided on the chalkboard is in fact a different form of uranium oxide known as triuranium octoxide (U_3O_8). This is also a real compound, to which the name uranium oxide could also be attributed. The formula provided, U_3O_8, is one of the most kinetically and thermodynamically stable forms of uranium. The remaining formulae and names are all correct. Marie focuses her attention on the "other extraneous matter" with a monologue stating that "there are 78 elements... and we believe that there are some elements still unknown". She declares "Pierre, we have discovered a new element, an active element". The accuracy of the onscreen chemistry throughout these scenes is commendable, which includes the correct number of elements known at the time. However, the only complaint is that the units for their measurements on the electrometer are not mentioned, and are only referred to as "points". In reality, they are most likely millivolts or milliamps, depending on what they are measuring.

Later, while trying to convince the university to fund their research into the isolation of radium, there are questions about Marie's abilities as a researcher because she is a woman. However, it should be noted that the funding board also worry about her inexperience. Pierre defends and vouches for her, but the budget provided by the university only allows for an old shed to use as their lab. The subsequent voiceover provides extensive details of their hard work: "They carried on between them the work of an entire chemical plant. The raw material was pitchblende, from the mines of Bohemia, tonnes of it, from which they planned to extract all the known elements until only a few ounces remained. From these few ounces, radium, their precious element, was eventually to be isolated".

The large-scale extractions eventually turn to bench-scale chemistry, with various forms of glassware used to separate the previously undiscovered radium from the known element barium. They both persevere through numerous failures, leading Marie to visit a doctor about an irritation on her hands. The doctor exclaims that he has never seen "burns quite like these before" and warns that the burns may develop into cancer if she continues her work. This is one of the few references to radiation burns during the movie, which contrasts sharply with that seen later in *Radioactive* (2019). Finally, they isolate radium, with the narrator explaining the process of crystallisation and recrystallisation with the image of hundreds of small evaporating bowls providing some sense of the scale of the work.

In the latter stages of the movie, the pair are visited by another famous chemist, Lord Kelvin from Belfast, who is depicted with an interesting attempt at a Northern Irish accent. But this could be forgiven since Kelvin was a travelled man at this stage of his career, so his accent would have changed over time. Lord Kelvin declares that he would love to stay and see the final crystallisation of radium, "but it's New Year's Eve and my family are expecting me in London". This is another reference to the importance of maintaining a personal life alongside the perseverance needed for scientific research. Contrasting this, Marie and

Pierre sleep in the lab that night, only to wake up and assume there is nothing in the evaporation dish since it looks empty. In this moment it appears that their obsession and perseverance does not pay off, despite all the sacrifices. However, once it turns dark, they can see a glow from the radium in the evaporating dish.

Earlier, while trying to convince the university to fund their research, Marie states that "we have carried on our work intensely, except for 5 weeks in the autumn. In September my daughter was born, and a week later my husband had the misfortune to lose his mother. But the rest of the time we have devoted entirely to research". This passionate speech illustrates the perseverance that the Curies gave to their work, which was ultimately successful. But, like the fictional movies mentioned earlier, there are also unforeseen side effects in the form of radiation poisoning. The effects of radiation poisoning are briefly mentioned and not explored in any great detail here, probably because it would be another three years before the first atomic bomb was developed and used, bringing with it the full horrors of radiation poisoning.

Overall, *Madame Curie* (1943) highlights the need for a healthy work/life balance, as well as the benefits of family and a break from science when things get difficult. After their success, the movie shows the Curies taking a holiday to recuperate. However, it also celebrates perseverance, like that shown in the other biographies from this era like *The Story of Louis Pasteur* (1936), *Edison, the Man* (1940), *Dr. Ehrlich's Magic Bullet* (1940) and *The Great Moment* (1944). We will return to Marie Curie again later in the modernised version of her story in *Radioactive* (2019). However, also worth a brief mention here are two examples where Marie Curie is depicted on screen using time travel. The first is the children's TV show *A.J.'s Time Travellers* (1994–1995), in the episode *Marie Curie* (1995). We also briefly meet Marie and Irene Curie in their mobile X-ray van on the battlefield in the TV series *Timeless* (2016–2018). In a brief sequence, they are requested to produce an X-ray image, but it turns out blurry, which is later blamed on the mothership that was used to travel back in time.

REFERENCES

1. M. Rose and H. Hogan, *A History of the Laser: 1960–2019*, Photonics Spectra, https://www.photonics.com/Articles/A_History_of_the_Laser_1960_-_2019/a42279 (accessed 2024-02-20).
2. S. Matskeplishvili, Vladimir Petrovich Demikhov (1916–1998), *Eur. Heart J.*, 2017, **38**(46), 3406–3410.

War and Peace

4.1 LIGHTS, CAMERA, ACTINIDES

As previously mentioned, tight budgets and travel restrictions limited movie making during the 1940s due to WW2. However, behind closed doors at the Manhattan project and through subsequent work, new chemical elements were being discovered during the development of the first atomic bombs. These new elements would later join the naturally occurring actinium, thorium, and uranium to become known as the actinides. We saw from our discussion about *Madame Curie* (1943) that it primarily covered her work with Pierre and did not mention her pioneering work in radiology with her daughter Irene through their mobile X-ray trucks on the battlefield during WW1. Instead, this part of her life is told in *Radioactive* (2019) and referenced through a brief cameo in the Canadian time-travelling series *Travellers* (2016–2018). This work represents the use of radiation for the good of humanity, contrasting with the use of actinides for war in atomic bombs. Here we will discuss onscreen chemistry examples with links to WW2, as well as the first post-war examples.

The first in our discussion here is another sequel to *The Invisible Man* (1933) called *The Invisible Agent* (1942), which tells the story of Frank Griffin Jr, grandson of the original *Invisible*

Onscreen Chemistry: The Portrayal of Chemical Science in Film and TV
By John O'Donoghue
© John O'Donoghue 2025
Published by the Royal Society of Chemistry, www.rsc.org

Man (1933). The movie opens with Frank hiding under a false name, but he is found by German agents who demand the "invisibility drug" that his grandfather invented. He manages to escape them and informs the US army about the incident, but it turns out the US army also want the drug. After initially denying their request and claiming that "there will never be an emergency critical enough to justify its use, never", he relents after the bombing of Pearl Harbour under the condition that he is the only one allowed to use it.

He is subsequently parachuted into Berlin to investigate the German plans to attack the US, using the drug to turn invisible to carry out the covert mission. Unfortunately, little chemistry of note is shown or discussed on screen, he is shown injecting himself on the plane and then turns invisible as he lands. Later he explains that he has a "certain chemical" in his veins which takes the colour out of his whole body. Interestingly, the side effect of the invisibility drug in this movie is that it makes him sleepy rather than mad, and the word 'drug' is mainly used rather than 'chemical formula' as in the previous movies.

Away from the world of supernatural science, one of the best examples of a war-themed chemistry movie is *The Adventures of Tartu* (1943), also called *Sabotage Agent* (1943) in the US. It follows an undercover British chemical engineer (Robert Donat) who accepts a dangerous assignment to steal the formula of a new Nazi poison gas and sabotage the chemical plant where it is being manufactured. However, upon arriving in Czechoslovakia, his contact is arrested and he is instead assigned to the Skoda munitions factory, which is a repurposed car factory. While protesting this decision, he exclaims "all my life I have worked at chemistry. It is not much, I ask not much to serve the Fuhrer in the best way I can. ... I am an expert, but an expert only at chemistry". It is certainly an impassioned show of pride for the chemistry profession, but it is initially unsuccessful.

Eventually, Tartu is transferred to the laboratories at the new chemical plant after proving himself at the munitions factory. His visit starts with the lead scientist asking Tartu if he has worked with the "metal in concentrates", to which he responds "Oh yes, for two and all three dihydro-cides". His response doesn't make much sense chemically speaking, unless it is an old reference to hydrogen cyanide (HCN), dihydrides, metal halides, or some play on words using the suffix-cide, meaning 'killing' or 'killer'. This suffix is used in chemical terms for pesticides and insecticides, but it is also used for homicide (the killing of one person by another) and regicide (the action of killing a monarch) among many other examples.

The laboratory is filled with glassware and various apparatus, including a precision scientific balance. A brief image of the handwritten chemical formula for the new poison gas is shown onscreen in a notebook. The following real chemical reactants are clearly shown: $KClO_3$ (potassium chlorate, used in fireworks), KCl (potassium chloride, used in low sodium table salt), P_2O_5 (phosphorus pentoxide, a strong drying agent), $C_7H_5(NO_2)_3$ (trinitrotoluene (TNT), an explosive) and O_2 (oxygen).

The resulting chemical structure of the proposed gas seems to have some chemistry errors, but it is difficult to make out the details. However, it looks like an organophosphine-type compound with a six-membered ring and two amine groups. In the real world, phosphines are highly toxic respiratory poisons and are usually associated with a foul smell. Organophosphines were first discovered in the 1930s and initially found use as pesticides. Overall, it seems that a lot of effort went into the creation of these notes, despite their brief appearance on screen. Also, the operation of use for the new gas is described as "we ship the gas to the various fronts in two separate liquid forms. There, the liquids are compounded, and the bomb casings charged just prior to operations". The use of 'binary chemical systems' which combine to create a dangerous substance will be used again later in *Die Hard with a Vengeance* (1995) and, more accurately, in *The Rock* (1996).

We are then shown a short montage of Tartu working in the lab wearing a lab coat with a mask and round safety goggles. He measures out some solid samples using a spatula and uses various forms of glassware, as well as a centrifuge, a microscope and other common lab equipment, as he prepares the gas. Tartu describes the gas as "inconceivable in its deadliness". He then contacts the resistance who give him access to their underground lab to create a "nitrocine bomb" (better known as nitroglycerin) to destroy the factory. Tartu is successful and becomes one of the first hero chemists depicted on screen, an archetype that we will discuss again in the next few sections and later in relation to *The Rock* (1996).

Nitroglycerin (also spelled nitro-glycerine) is a dense, colourless, oily, explosive liquid first discovered by the Italian chemist Ascanio Sobrero in 1847, who warned against using it as an explosive due to its instability. One of his students was Alfred Nobel, who continued to research the substance which eventually led to 'blasting oil'. Later, he also mixed it with inert substances to form 'dynamite', which was safer to handle. However, the shock sensitivity and instability, even at moderate temperatures, led to numerous accidents around the world, especially during the American railroad expansion of the 19th century.

In the UK, several accidents eventually led to the *Nitroglycerine Act* of 1869 which greatly restricted its use. Therefore, it shouldn't come as much of a surprise that it frequently finds its way into onscreen chemistry. However, it is also used in medicine as a potent vasodilator to treat heart conditions such as angina pectoris and chronic heart failure.[1] But this is rarely mentioned onscreen. The unusual shock-sensitive nature of the substance makes it a favourite for onscreen explosions, taking the mantle as the most popular explosive. It is even used as a plot device in a children's animated TV series called *Cadillacs and Dinosaurs* (1993–1994). The plot revolves around the shock sensitivity as they transport jars of it through the jungle to "blow the canyon", creating a 'fire wall' to stop the spread of a wildfire.

Less than a year after the Trinity nuclear test and the subsequent dropping of the atomic bombs on Japan, we have a 13-part serial called *Lost City of the Jungle* (1946), about a fictional element called meteorium "with an atomic weight of 245", described as "the only practical defence against the atomic bomb". The serial follows a "warmonger" called Sir Eric Hazarias who has traced the only meteorium deposit to the fictional Himalayan nation of Pendrang. Each episode opens with scrolling text exactly like that used later in the *Star Wars* franchise. There are also numerous scenes in a secret basement lab with an elaborate glassware setup designed for geological testing. Condensers, Bunsen burners, graduated cylinders, conical flasks, and test-tubes are all easily identifiable. The scientist who discovers the element isn't concerned about the profits from it or what Sir Eric wants with it, he is only motivated by fame and for "the great scientists of the world to bow to the discoverer".

Meteorium is described as highly radioactive and "the most powerful element in the world". Eventually they find it locked in a chest in an ancient temple, where it is described by locals as the "eternal sun". Only robes made from asbestos by the local elders can be used to look at it safely. Later, Sir Eric escapes with the meteorium, but the scientist reveals that he sabotaged the case to explode to prevent it from getting into the wrong hands. Meteorium was likely inspired by the real-world discovery of curium, which has an isotope with an atomic weight of 245. Curium was discovered alongside americium in 1944, but the discovery of both elements was kept confidential because they were related to the Manhattan Project. Glenn Seaborg accidentally leaked the discovery of both elements on a US radio show for children, five days before the official announcement in 1945, when a listener asked if any new transuranic element beside plutonium and neptunium had been discovered during the war.

Finally, worth a brief mention here is *The Third Man* (1949), about an author who travels to post-WW2 Vienna, which has been divided among the allies like Berlin. He intends to meet his childhood friend who has offered him work. Instead, he stumbles on a conspiracy involving the death of his friend who was stealing penicillin from the laboratories in a military hospital. His friend was then "diluting the penicillin" and selling it on the black market to those who needed it. However, because it wasn't strong enough in its diluted form, those with gangrene and meningitis were not able to recover properly.

4.2 CHEMISTRY CRIMES

Long before the *CSI: Crime Scene Investigation* (2000–present) franchise brought forensic science to the small screen; *Kid Glove Killer* (1942) brought the professionalisation of forensic chemistry to the big screen. The story begins when a new mayor promises to clean up crime in the city, which prompts an organised crime outfit to commit a prominent murder. The "scientific crime laboratory" opens with a "police chemist" called Gordan McKay (Van Heflin) looking through a microscope. As the camera zooms out, it reveals a well-stocked and comprehensive lab filled with several shelves of chemicals. He calls for his female colleague named Mitchell (Marsha Hunt), who is using some glassware and a Bunsen burner on the other side of the lab. After mixing some liquids and holding up a test-tube, Mitchell declares "no trace of cyanide", but he questions her, "did you rule out the alkaloids?", to which Mitchell firmly responds "Now look Mr McKay, I hate this job and you may think I'm a creep at it, but I do know how to make an analysis".

Mitchell then reminds him that she has a master's degree in chemistry. They then proceed to have a cigarette together, which they light using a Bunsen burner in the lab. The wise-cracking dialogue continues when the prosecutor enters to ask for an update on the murder, calling McKay "Sherlock". When asked about his new female colleague, McKay cracks that the lab was getting dirty, so he needed a maid to clean it up. He then claims that she "hates chemistry" and the prosecutor responds, "she should!". Throughout the murder investigation, various real substances are mentioned such as tannic acid, cresol and iodine spray among others. Also, McKay and Mitchell aren't confined to the lab, they visit suspects, carry out field work, and even fight off thugs with guns, like the modern iterations of *CSI*. (Figure 4.1)

Later, Mitchell reveals that she does really hate chemistry. When she is asked "why do you stick at it then?", she responds "I had a horrible childhood; it taught me to eat". The prosecutor then quips "chemistry or starve – not for the rest of your life, I hope". She then tells him that she's only sticking with it until she gets married. The remainder of the movie centres around solving the murder of the mayor by analysing the fragments and remains from a bomb detonated under his car. They use a spectrograph to analyse the powder, which they explain produces "a spectrum, a series of tiny vertical lines". Using a line spectrum, McKay goes on to explain that it can be used to match up which chemical elements are in the powder. The spectrum reveals that the elements are mainly potassium and vanadium. Then, using a reference book, the words "nitrocellulose", "barium nitrate", "vanadium" and "potassium" are shown on screen as the main contents for that type of explosive.

Noteworthy are the frequent comments from McKay throughout about "keeping the wolf from the door" and being underpaid. He proclaims that forensics isn't a career for a man

Figure 4.1   An early depiction of a forensics lab in *Kid Glove Killer* (1942), featuring a police chemist and his lab assistant. The tag line for the movie is "crime under the microscope". © Moviestore Collection Ltd/Alamy Stock Photo.

or a woman since there is "no prestige, no glamour, no money," and "people holler at you when there are no miracles". Despite this, the forensic chemists stay professional even when they become deeply conflicted about who the real killer is. Both of the chemists are presented as the good guys and turn out to be uncorruptible, unlike many other characters in the movie. McKay also emphasises the need for multiple pieces of evidence before giving a theory and they should not jump to conclusions. Overall, *Kid Glove Killer* (1942) is smart, accurate, and humorous, standing as one of the first depictions of forensic chemistry long before the more familiar modern versions dominated this genre.

Worthy of note here as well is *D.O.A* (1949), which features a reference to toxicology. The onscreen murder involves a "luminous toxin", but the details of what the toxin might contain are not provided, only that it is a "poison that attacks the vital organs". There is also a brief scene in which a doctor shows a glowing test-tube, demonstrating the luminosity. The movie makers probably used luminol for this scene, which was first discovered in 1928 by the German chemist H. O. Albrecht.[2] It later transpires that the victim was murdered for notarising the bill of sale for iridium, a rare chemical element that will appear again in relation to *The Martian* (2015). It is normally mixed with platinum for jewellery and used in nibs for pens due to its hardness, but its use in this movie is not clear. Luminol will also appear again several times since, among other uses, it is used in the real world to detect blood at crime scenes.[3]

Finally, we will finish this section with the British movie *Thunder on the Hill* (1951), based on the play *Bonaventure*. During a storm, a flood results in road closures, forcing a police officer to bring a murderer called Valerie to a convent hospital to wait out the storm. Valerie is accused of giving her brother an overdose of medicine and most likely faces the death penalty. It includes some scenes where the doctors and nuns are working in a lab preparation room, referred to as a medicine cupboard, with a large array of glass containers. However, other than a brief mention of penicillin in an early scene, the medicine involved in the overdose is never mentioned. Also, when Valerie declares her innocence, she states that she did not substitute the aspirin for extra doses of the unnamed medicine. Despite the limited chemistry references, *Thunder on the Hill* (1951) stands an interesting early example of death by overdose rather than *via* a poison.

4.3   ETHICS AND CHEMISTRY

After WW2, scientists like J. Robert Oppenheimer became famous due to their efforts to stop the Axis powers with developments like the atomic bomb, which will be discussed again later.[4] By the end of the 1940s, the world had entered the atomic age. Rebuilding after the war kickstarted an expansion in scientific industrialisation and agriculture unlike anything previously seen. Before the war, chemical industrialisation was depicted as a debate between the benefits and fear of progress, as shown in *Things to Come* (1936). During the war, we saw how the chemical industry was hijacked by the Nazis to produce poisonous weapons like that seen in *The Adventures of Tartu* (1943).

However, towards the end of the war, industrial ethics begins to rise to prominence. This theme will evolve over the coming decades, resulting in well-known movies like *Erin Brockovich* (2000) and *Dark Waters* (2019), among many others. The first example here is a movie called *I Know Where I'm Going* (1945), where the protagonist announces to her bank-manager father that she is engaged to the chairman of CCI (Consolidated Chemical Industries). Her fiancé, Sir Robert Bellinger OBE, is described as "one of the richest men in England" and requests that she join him on a remote island in the Scottish Hebrides where "the war is a million miles away".

While on the overnight train to Glasgow, she dreams that she is marrying the company (CCI) rather than the person (Bellinger). In her dream she is asked if she will take Consolidated Chemical Industries as her lawful wedded husband and *vice versa*. Bellinger is never seen on screen, only heard through voiceovers and on the radio. He is frequently called the "rich man on the island" by the locals, and he is said to be supplying "chemicals for the war effort" while hiding far away from the war. It depicts him as out of touch with normal life due to the vast wealth that he has accumulated from chemicals. The name of the fictional Consolidated Chemical Industries (CCI) has a striking resemblance to the real-world Imperial Chemical Industries (ICI), who were rapidly expanding their operations at the time and generating enormous revenue in the process.

A few years later we have another movie from this era with unethical decisions in the name of business called *Miracle in Harlem* (1948). The story follows a veteran from the "Chemical Warfare Unit" who is described as a "laboratory man". The plot centres around the use of poison to dispose of the new owner of a small candy business after he swindles the business away from an elderly lady. Of note is a statement during the movie that "a working knowledge of a laboratory is essential for good candy making". *Miracle in Harlem* (1948) is very commendable and noteworthy from a diversity point of view, depicting multiple black chemists on screen. However, the chemistry references are vague, existing for the most part only to move the plot along.

The following year we have *Riders of the Whistling Pines* (1949), the first western with a chemical theme. It begins when a forestry company notices a tussock moth infestation in the trees, but they keep it to themselves so they can harvest more timber outside their designated area. Later, a former ranger also discovers the infestation and, despite an attempt by the forestry company to stop him, reports it. In the next scene, some officials conclude that the "only answer is to spray the entire area with the DDT solution" (Figure 4.2). They ask the local doctor whether he thinks "the poison will affect wildlife or domestic stock up there", to which he responds "No, I don't think so, if used in proper proportion".

Considering this, the foresters begin a local campaign against DDT by trying to convince farmers that it will kill off their livestock, thus creating an interesting dichotomy between the unethical forest harvesters and the potentially dangerous DDT. When the trucks arrive, the foresters start their misinformation campaign with statements like "I hate to think what will happen when they turn it loose from those airplanes", to which a local responds "they claim it won't do any harm". The forester goes on

Figure 4.2   Spraying crops with dichlorodiphenyltrichloroethane (DDT) after WW2, as referenced in the movie *Riders of the Whistling Pines* (1949). CDC/Science Photo Library.

to say "it'll kill the bugs in the trees, it'll kill everything else won't it?". Soon enough, a second-hand unchecked story emerges: "I heard of a fella who sprayed the DDT on a dog to kill the fleas. It killed them alright, but it killed the dog too". Another local then asks "what's going to happen to the fish in the steams?", but she is dismissed as talking foolish because the timber is more important.

Riders of the Whistling Pines (1949) is a fantastic exploration of misinformation and fear-mongering based on incomplete information. The full name of DDT is never given in the movie and the foresters are depicted as the antagonists, but the fear they propagate about DDT turns out to be well founded in real life. In the movie, dead animals like deer and foxes are found near a stream, which leads to an interesting discussion on screen about whether DDT was properly tested and whether the correct dose was used.

The lead character continually defends the use of DDT, claiming it was properly tested and the dose was correct, echoing the official information available at the time. It later transpires that the foresters were spraying something "stronger" to kill the wildlife. In a brief scene, the words "OPERATION FLIT GUN" is seen on the side of a tanker truck. FLIT is a real-world brand name for a popular insecticide used between 1928 and 1950, and a FLIT gun is a manual spray pump used for atomising it. FLIT also contained DDT and the name has since been reused as a brand name for modern insecticides (without DDT).

Dichlorodiphenyltrichloroethane (DDT) is a colourless, tasteless, and almost odourless compound whose insecticidal properties were discovered by the Swiss chemist Paul Hermann Muller in 1939, who won the Nobel Prize in Physiology or Medicine in 1948 for his work. It was first used during WW2 to stop the spread of malaria, typhoid and typhus among troops. It became available for public use in the US in 1945, with the World Health Organisation (WHO) and others encouraging its use throughout the 1950s and 60s to stop the spread of malaria. However, from the very beginning there were articles and advertisements urging caution when handling DDT. By the late 1940s there were already movements against DDT, with some convinced that it was making Americans sick and that it was killing off chicks and bees.[5]

Interestingly, opposition to DDT did not pick up pace until the publication of Rachel Carson's book *Silent Spring* in 1962. *Riders of the Whistling Pines* (1949) was released long before this, making it the first mainstream movie to highlight the potential dangers of the insecticide, albeit by unethical foresters trying to make a profit from timber cutting. It was well known by the end of the war that DDT could have detrimental effects on life, with warnings from the US War Department not to spray it on cattle, fowl, and fish or on "waters that might be used for human consumption". DDT was eventually banned for agricultural use in the United Sates in 1972 and later worldwide in 2004.[2]

4.4 CHEMISTRY INVENTIONS

Over a decade after *The Love Test* (1935), we return to a female research chemist in a lead role in *Strange Impersonation* (1946), which tells the story of Nora (Brenda Marshall), who invents a new anaesthetic. The movie opens in the "Wilmott Institute, Chemical Research" with Nora confidently presenting her research and identifying the part of the brain that her proposed anaesthetic will target. She claims it will last one hour, during which time the mind will engage in dreams or fantasies. As she finishes, Nora packs up some lab glassware including volumetric flasks and graduated cylinders. The next scene brings us into the lab which contains an elaborate setup of glassware complete with burettes, beakers, condensers, and chemical bottles.

Nora decides to test the new anaesthetic on herself in her apartment, outside the control of the institute, despite pleas from her lab colleague to instead do it in the lab where they have oxygen. Nora then asks her colleague to check on the distillation at the other side of the lab before her love interest enters the room. After he leaves, Nora is asked when they are going to marry, to which she replies that they haven't decided. Her colleague then muses "Oh, I was wondering if I like working in the lab so much that I'd let it delay my marriage and I don't think so", most likely referencing the 'marriage bar' which was still in operation in many counties at the time. This required women to resign from their job once married.

When Nora heads home for the evening she nearly runs over a pedestrian in her car. Her life then begins to unravel after the new anaesthetic explodes in her apartment, resulting in her friend stealing her fiancé while she recovers in hospital. Later, we are treated to a few more lab scenes with plenty of glassware filled with solutions, as well as laboratory techniques such as pipetting. Finally, it transpires that most of the movie was a dream induced by the success of her anaesthetic, but the experience convinces her to embrace marriage after all. Despite the association of chemistry with explosions, *Strange Impersonation* (1946) builds on the transition of fictional chemistry on screen, echoing the accurate chemistry portrayed in biographies from this period as discussed in *The Great Moment* (1940).

The post-WW2 world saw a plethora of new inventions in the real world such as DDT, the microwave, colour television, Velcro, diving suits and so on. So, it should come as no surprise that movie makers also jumped on the inventor bandwagon to have some fun with fictional inventions of their own. *It Happens Every Spring* (1949) is about a chemistry professor who invents a new compound which he finds useful in baseball, a literal interpretation of the 'screwball comedy' genre. From the very beginning, chemistry, labs, and PhDs are discussed in relation to the lead character, named Vernon (Ray Milland). It is said that he has taken longer than planned to complete his PhD because he spent three years in the South Pacific in the Chemical Warfare Unit during the war. However, it is also mentioned that every spring he gets very distracted and aloof, around the time the baseball season starts.

While teaching a lesson on acids, various test-tubes and storage flasks are shown on his bench. Behind him on the board is the correct equation for the deprotonation of carboxylic acids in the presence of a base, forming H_2O. The chemical formulae for various carboxylic acids are also shown on the other side of the board along with their K_a (acid dissociation constant) values, all of which are correct. He proceeds to carry out a demonstration using acetic acid and methyl orange, which he describes should turn pink but gets distracted by the baseball game on the radio under his desk.

Vernon later explains that his "nitrocyclohexane" compound might have great commercial value as a "bio-phobic", a substance that keeps insects away from wood *etc.*[†] He says that this will finally provide him with the money to ask his girlfriend to marry him, which is a similar argument to the one used earlier in *The Invisible Man* (1933). The lab is shown with an array of condensers and large flasks containing bubbling liquids and white condensate similar in appearance to CO_2 dry ice, which becomes a standard bearer for chemistry on screen by this stage. He shouts in delight "Hallelujah! There it is, the white precipitate", which is shown in a flask. As he dutifully records his results in his lab notebook, a baseball flies through the window and smashes his entire experimental setup. He is distraught, exclaiming "one of those compounds alone took five weeks to crystalise and they're all in sequence. I can't make the second until I finish the first".

However, as he begins to clean up the mess, he discovers the baseball landed in the precipitate and now repels wood. He recruits two baseball players from the college team to test his discovery and in return he gives them pointers about organic chemistry. He tells them they are having trouble with the basic terms and begins naming "methyl, ethyl, propyl". He proceeds to run the experiment with the students by adding a cloth with his compound into his glove: "You hit three consecutive pitches and then you missed three in a row. Statistically, therefore, I have obtained all the information possible. Unless, of course, I pitch several hundred more balls". Vernon then exploits his invention as a pitcher for the St. Louis baseball team under the false name of Kelly and keeps the compound a secret.[‡]

However, he starts to feel guilty, and the university also begins searching for him since he didn't turn up at the industry labs,

---

[†] Nitrocyclohexane is a real organic compound with the formula $C_6H_{11}NO_2$, and is a colourless, highly flammable liquid and strong oxidising agent. It is generally used in the production of various other chemicals.

[‡] The tune to the song *Has Anyone Here Seen Kelly* is frequently played in the background throughout the movie. The song was originally entitled *Kelly from the Isle of Man* and was written for late-Victorian music halls in 1908. It's about a Manx woman looking for her boyfriend during a visit to London. It was later adapted for American audiences by Nora Bayes, switching the Isle of Man for Ireland and London for New York City. The American version briefly appears in *Catch Me If You Can* (2002) when Leonardo DiCaprio's character is watching a real-world 1960s TV show (*Sing Along with Mitch*) with his fiancée's family.

where he claimed he was going. He sends a letter and a diamond ring home explaining that he isn't missing, but the financial rewards are "quite fantastic". This immediately prompts suspicion, "Isn't that rather unusual for scientific demonstrations?", to which the dean responds "it's not unusual – whatever he is doing is not legitimate". Eventually he runs out of his formula before the last game, but he wins honestly by catching the strike, breaking his hand in the process, and finishing his baseball career. He concludes by saying "A lot of things don't make sense. I was a chemistry teacher, I can tell you that now. The sum of money I received for teaching science to the youth of this state for an entire year was a little less than I got in a single afternoon for tossing a 5 oz sphere past a young man holding a stick". He then returns home to a huge welcome.

Following this, we have another chemistry-related invention in *Yellow Cab Man* (1950), a screwball comedy about a clumsy, accident-prone inventor who creates "elastiglass". As the plot develops, unscrupulous legal firms attempt to steal his formula for the glass because they fear that it will reduce the number of accident claims. Towards the end of the movie, they use a form of induced hypnosis which results in a flash back scene where he remembers experimenting in his basement using "an excellent book on chemistry".

The next scene shows a home-made setup of metal pots with tubes, and he says he wanted to follow in the footsteps of Edison, Pasteur, Madame Curie, and the Wizard of Oz. We then move to a scene in a large lab filled with glassware and he mentions that he was experimenting with silica, lime, potash, and lead oxide. All of these are real chemical compounds used in the manufacturing of glass. After an explosion he finishes by saying "the final step was in the fusing of the silica with the alkali and elastiglass was born". Overall, *Yellow Cab Man* (1950) is a comedic interpretation of the real-world invention of laminated glass in 1903 and Plexiglas in 1933, but the chemistry content is unfortunately very limited.[6]

In contrast, just a year later, we have a classic of onscreen chemistry called *The Man in the White Suit* (1951). This British satirical comedy is about a self-absorbed chemist who invents a fabric which can resist wear and stains, meaning it never needs to be cleaned. However, big business and labour unions realise it must be suppressed for economic reasons. A fabric which never wears out would result in reduced sales, and if it never needs to be cleaned, it will put cleaning services out of business too. The fabric eventually unites the entire textile industry, from the labourers to the researchers to executives because they all fear that this invention will put them all out of a job.

The story revolves around Sidney Stratton (Alec Guinness), a brilliant young research chemist and former Cambridge scholarship recipient who is interested in polymer chemistry. He is frequently described as a "lunatic" throughout the movie due to his obsession with the research and his selfish nature, frequently using those around him to further his obsession. The story is loosely based on the real-world discovery of nylon by Wallace Hume Carothers at DuPont's research facility in 1935. Nylon was one of the first materials engineered from scratch, based on an understanding of polymer chemistry and a desire to plug what was a serious hole in the hosiery department.[7,8]

The movie opens in a chemistry lab filled with glassware, bubbling solutions, melodic sounds, and chemists in lab coats. Our protagonist Sidney is working on a side project within the lab which attracts the interest of Mr Birnley who is conducting a tour. As they try to figure out who owns the elaborate setup of glassware, a chemist is distracted from mouth pipetting a solution which causes him to inhale the liquid. This was a common technique at the time but is no longer acceptable in modern labs and has been replaced with pipette fillers. The experimental setup resembles some form of distillation, albeit spruced up with some flashing lights and melodic bubbling sounds. We later discover that he spent £4000 of the company's money on "heavy hydrogen" (also known as deuterium) without permission.

We also discover that he has been dismissed from several other jobs at textile mills due to his demands for expensive equipment and his obsession with inventing an everlasting fibre. Later, while working as a labourer at the Birnley Mills, he is eventually assigned a task to deliver a box to the chemistry lab which is presented with great triumph. While neglecting his duties as a labourer, a page from his textbook is briefly shown on screen showing the chemical structure of a polymer. However, it appears to contain an impossible triple-valent oxygen atom nestled between six- and five-membered rings that resemble carbohydrates. Eventually, he gains access to the lab when he is asked to bring up a new electron microscope which has just been delivered.

While helping the research team set up the microscope, he accidently becomes an unpaid researcher when they mistake him as a technician from the manufacturer. This allows him to continue his work on the fabled everlasting fabric. We later discover that Sidney also has a chemistry setup in his bedroom and his obsession confuses all of those around him. They do not understand his desire to work in the lab for no money with no means of paying for food, leading his colleagues to think that he is in some kind of trouble. This echoes the theme of obsession that was common in movies before the war. Usually, the obsession with scientific research can be harmful to one's health and wellbeing and can also ultimately lead to the character's downfall. However, his bedroom lab setup is a far cry from the elaborate home labs seen previously, setting Sidney apart from the established elitist image of chemistry. Despite this, a familiar phrase from *Things to Come* (1936) also reappears towards the end of the movie when his landlady exclaims "Why can't you scientists leave things alone?".

His covert activities are eventually discovered by Birnley's daughter, Ms Birnley (Joan Greenwood). In a fantastic example of simplifying the chemistry, Ms Birnley's character plays the role of the audience as Sidney attempts to explain his work part-by-part: "Look, you know about the problem of polymerising amino acid residues? You know what a long chain molecule is? You know what a molecule is?". She figures out that a molecule is "something like an atom", which Sidney latches on to so he can explain a polymer as "atoms stuck together, in this case like a long chain". He gives cotton and silk as natural examples as well as Rayon and Nylon as artificial examples with even longer chains.

Despite blowing up the lab numerous times, he eventually creates the titular white suit from his newly discovered fabric through the "co-polymerisation of amino acid residues in carbohydrate molecules, those containing ionic groups". This description would imply, from a chemical viewpoint, that his fabric resembles glycoproteins or sugars. Interestingly, during his excitement he mumbles that he thought the polymerisation would be "sterically hindered", which is a molecular property where bulky components physically block others from interacting. The suit is white because it cannot absorb dye, but it is also luminous because it contains "radioactive thorium". Sidney impresses Ms Birnley so much that she is encouraged to study chemistry from the *Encyclopaedia Britannica* so she can understand what he is talking about. Later, she then uses her newfound knowledge to describe Sidney's work to her father to convince him to fund the work.

The Man in the White Suit (1951) stands as a classic of onscreen chemistry, with a large percentage of screentime dedicated to chemistry labs and glassware. Despite the fictional chemistry involved, it provides us with an accurate look at real chemistry skills in an industrial research lab. It also explores the tensions between the established order of the pre-war industrial world (Mr Birnley) and the new age emerging after the war, full of scientific advancement led by a younger generation (Sidney and Ms Birnley). However, the protagonist is depicted as a one-dimensional chemist, with no interests, hobbies, relationships, or life outside of his obsession with trying to invent the fabric. Even his bedroom contains experimental equipment and books, echoing previous depictions of obsessive one-dimensional chemists. Real-world chemists are of course significantly more complex, and, one would hope, are also capable of recognising the broader implications of their work.

Interestingly, *The Man in the White Suit* (1951) isn't the last time Joan Greenwood plays a character in a movie which includes laboratory glassware. *Mysterious Island* (1961) is based on the Jules Verne novel of the same name from 1874 and continues the story of Captain Nemo from *Twenty Thousand Leagues Under the Sea* (1954). In the latter part of *Mysterious Island* (1961), large conical flasks are used by Captain Nemo to demonstrate and explain displacement. They are also used again in the closing stages of the movie by the other captain to show how a balloon can raise a ship from the seabed so they can escape.

REFERENCES

1. N. Marsh and A. Marsh, A Short History of Nitroglycerine and Nitric Oxide in Pharmacology and Physiology, *Clin. Exp. Pharmacol. Physiol.*, 2000, 27(4), 313–319.

2. H. O. Albrecht, Über Die Chemiluminescenz Des Aminophthalsäurehydrazids (On the Chemiluminescence of Aminophthalic Acid Hydrazide), *Z. Phys. Chem.*, 1928, **136**, 321–330.
3. W. Specht, Die Chemiluminescenz Des Hämins, Ein Hilfsmittel Zur Auffindung Und Erkennung Forensisch Wichtiger Blutspuren (The Chemiluminescence of Haemin, an Aid to the Finding and Recognition of Forensically Significant Blood Traces), *Angew. Chem.*, 1937, **50**(8), 155–157.
4. K. Bird and M. J. Sherwin, *American Prometheus: The Triumph and Tragedy of J. Robert Oppenheimer*, A. A. Knopf, New York, 2005.
5. E. Conis, Beyond Silent Spring: An Alternate History of DDT, Distillations Magazine, 2017, https://www.sciencehistory.org/stories/magazine/beyond-silent-spring-an-alternate-history-of-ddt/.
6. Plexiglass, *THIS IS HOW PLEXIGLAS® CAME TO BE*. https://www.plexiglas.de/en/about-us/history.
7. Royal Society of Chemistry, *Making nylon: the 'nylon rope trick,'* RSC Education, https://edu.rsc.org/experiments/making-nylon-the-nylon-rope-trick/755.article.
8. A. J. Wolfe, Nylon: A Revolution in Textiles, *Distillations Magazine*, 2008, https://www.sciencehistory.org/stories/magazine/nylon-a-revolution-in-textiles/.

The Golden Era

5.1 ALL THE BEST JOKES ARGON

The 1950s represented a boom in movie production after WW2, giving us numerous well-known classics such as *Cinderella* (1950), *A Streetcar Named Desire* (1951) and *Singin' in the Rain* (1952), to name just a few. Budgets increased and equipment was finally upgraded to facilitate colour movies. Following our previous discussion about the silent era, *Singin' in the Rain* (1952) explores the friction between stage and movie actors, as well as the physical problems of recording sound for the early talkies. It centres around the squeaky voices of some actors and actresses who managed to hide their flaws during the silent era but were exposed when sound was introduced. It also highlighted the need for screen writers, since early onscreen dialogue was found to be repetitive and dull.

The previous season finished with a discussion about wacky chemistry inventions in *The Yellow Man* (1950), *It Happens Every Spring* (1949) and *The Man in the White Suit* (1951). We also saw how the screwball comedy genre was becoming popular for portraying chemistry and chemists onscreen. Following this trend, we have America's answer to the British-made *The Man in the White Suit* (1951) with another screwball comedy set in an industrial chemistry lab about a fictional chemistry invention.

Onscreen Chemistry: The Portrayal of Chemical Science in Film and TV
By John O'Donoghue
© John O'Donoghue 2025
Published by the Royal Society of Chemistry, www.rsc.org

Monkey Business (1952), starring Cary Grant and Marilyn Monroe, among others, follows an absent-minded and frequently distracted research chemist called Barnaby who is trying to develop a "formula" for the elixir of youth. He wears thick circular glasses, a throwback to the lab glasses used in early 'mad scientist' tropes like Van Helsing in *Dracula* (1931).

His absent-mindedness is linked to the fact that he is thinking so deeply about his research, similar to the depiction of Pierre in *Madame Curie* (1943). The movie opens with Barnaby exclaiming "that's the trouble about being a chemist, you can't actually think. Every now and then you feel compelled to sit and stare at a sheet of white paper hoping it will come to you, but it never does". When Barnaby and his wife decide to skip a party to stay home together instead, their friend visits the house to investigate their whereabouts and states "Oh, the genius is at work again. I'm glad I'm a lawyer and not a chemist".

Following an emerging theme in the late 1940s, the well-stocked research lab once again contains animal test subjects like those discussed previously. Barnaby is shown pouring a liquid from a measuring cylinder into a beaker and then adding a powder while stating what sounds like "sodium acaptate" which is most likely a mispronunciation of the real chemical substance sodium acetate. His colleague arrives at his workbench carrying what he calls "molybdenum" to which Barnaby responds "sodium molybdate?". Sodium molybdate is a real compound (Na_2MoO_4), normally used as a fertilizer and as a corrosion inhibitor. He also hands Barnaby another powder but it is unclear what he calls it.

Later, a chimpanzee gets loose in the lab and mixes some liquids into a beaker from measuring cylinders and volumetric flasks after watching Barnaby working earlier. Perhaps this is a good example of why it's important to always tidy one's lab bench! The well-trained chimpanzee is probably the first example of an animal doing chemistry onscreen, and it's better than many human onscreen chemistry examples up to this point. The chimpanzee then dumps the concoction into the watercooler, setting up the antics for the movie. Barnaby also self-experiments in the lab using his new formula, despite a warning from his colleague that "self-experimentation is against the rules of all good research". This echoes a similar scene and statement from *Strange Impersonation* (1946), as discussed previously. In response, Barnaby claims "the history of discovery is the history of people who didn't follow rules" and proceeds to wash it down with water from the cooler, which, unknown to him, contains the actual youth formula.

The rest of the movie mainly pokes fun at the clothes, haircuts, cars, humour, emotional state and mannerisms of young people in the 1950s as Barnaby and others begin to feel younger. The formula wears off over time, which is a possible reference to metabolism as the body breaks down the substance. Due to a series of mishaps related to the undesired effects of "youthful behaviour", Barnaby decides that he wants to destroy the formula, ripping up his notes. They proceed to drink coffee from beakers, which was made using water from the cooler, prompting a repeat of the antics from earlier. Overall, *Monkey Business* (1952) once again references the 'research is dangerous' theme. However, in contrast to earlier examples, Barnaby and his wife are happier at the end, having learned to appreciate maturity and each other. Also, whenever he misses dinner and social occasions due to his obsession with work, he spends the time with his wife instead. He also works as part of a team, unlike Dr Jekyll who is always working alone in his lab.

5.2   TEACHING CHEMISTRY

The following year we move from the industrial chemistry lab to the university chemistry teaching lab with *The Affairs of Dobie Gillis* (1953), another screwball comedy. It starts off with the titular character Dobie (Bobby Van) expressing more interest in meeting women than learning. He meets Pansy (Debbie Reynolds) who is only interested in learning and working. To impress her, Dobie decides to take the same classes as Pansy, which includes chemistry. In the chemistry lab, we see a blackboard in the background with the chemical equation for the reaction of sulfuric acid (H_2SO_4) with zinc (Zn), which includes their correct "equivalent weights". We can also see the correct chemical formula for potassium chlorate ($KClO_3$) and potassium chloride (KCl), as well as Avogadro's number (6×10^{23}) and the equations for the decomposition of water. The professor describes the course as "the hardest course you've ever taken in your lives", which isn't exactly a great selling point for chemistry education.

The first experiment that they need to conduct involves the "decomposition of potassium chlorate", presented to them in a tube which they are told is connected to a water trough. They are told to use their Bunsen burners to gently warm the potassium chlorate, producing a gas which bubbles through the water into the gas jars. Pansy is excited by the work, gleefully exclaiming that "the chemist's life is a busy one". Her excitement is metaphorically linked with the experiment, causing the bubbles to fill her gas jar faster than anyone else's. The more excited she gets, the more bubbles are produced, until, off screen, there is an explosion and a flash of light. Presumably this implies that she has burst with joy due to her increased entropy and since they are linked, the experiment also exploded.

Although this is a perfectly real experiment, it is unfortunate that it is never explained what gas is being collected, which is oxygen. So, in theory, an explosion is possible if the Bunsen burner were to come into contact with the gas. Later, we are treated to a lab montage with brief images of other chemistry experiments. One involves a round-bottomed flask and a thermometer, but it is not clear what they need to do, while another involves the hydrolysis of water, which is reinforced by the correct equations on the chalk board. In all cases, the experiment explodes in line with Pansy's increasing excitement. After an absence of a few weeks, Dobie and Pansy return to the lab to discover that they need to figure out what their individual unknown samples are, a common university lab experiment still used today.

Because they skipped so many classes, they decide to pull an all-nighter by sneaking into the lab after hours to run the various tests and determine what their samples are. Pansy exclaims that she loves the lab work, but Dobie convinces her that they don't need any more explosions. As a result, Pansy reads out the notes for a metal test "add dilute hydrochloric acid with constant stirring until a precipitate ceases to form and decant the supernatant liquid". After a few tests, they finally determine that Pansy's sample is silver chloride. Dobbie then takes a nap and warns Pansy not to do "anything with the chemicals" in the meantime. However, as she empties the test-tubes into a large flask, she gets excited as the flask starts to smoke and foam, eventually causing a large explosion which destroys the entire lab and sets off the alarms and sprinklers.

Because the lab shown in *It Happens Every Spring* (1949) is a university research lab, *The Affairs of Dobie Gillis* (1953) may be the first chemistry teaching lab on screen. Despite the lack of safety goggles and reinforcing the "chemistry causes explosions" trope, the accuracy and attention to detail is commendable. Even the test-tubes from the unknown sample tests clearly contain precipitates as you would expect. The movie finishes with the chemistry lab exploding again, but Pansy is not involved this time, so apparently chemistry labs just spontaneously explode all the time. This is heavily contrasted with another movie featuring a chemistry teaching lab on screen a year later called the *The Belles of St. Trinian's* (1954), a British movie featuring a school chemistry lab.

The story here revolves around a sultan's daughter who is sent to a prestigious English "School for Young Ladies" in the same county where the sultan's racehorses are trained. Most of the movie is concerned with the horses, the inside scoop and winning the Gold Cup. The students are generally portrayed as loud, boisterous, mischievous, and unpredictable, with all the local townspeople hiding when the girls return from a hockey match. The chemistry lab only appears in a brief scene as the headmistress is showing the new teacher around the school. As the headmistress opens the door to the chemistry lab, she places a handkerchief over her mouth as a large cloud of smoke pours out.

First, we are shown a student smashing something in a mortar and pestle with a white powder spilling out all over the lab bench. The headmistress cautions her to be careful with that "nitroglycerin", but since nitroglycerin is an oily substance, it is not clear what she is referring to. This scene also contains a large round-bottomed flask bubbling with a funnel and a shelf of chemicals in the background. As they proceed further into the lab, they observe an elaborate setup of round-bottomed flasks, conical flasks, condensers and other forms of laboratory glassware all connected with tubing – a common feature of onscreen chemistry labs at this point, *e.g. Dr Jekyll and Mr Hyde* (1941) and *It Happens Every Spring* (1949).

The camera pans along the elaborate glassware and tubing apparatus where we see various liquids bubbling, eventually leading to what appears to be a distillation apparatus filling some bottles. When asked, the teacher, rather irresponsibly, admits that she hasn't a clue what they are making. It turns out to be gin, which they are packaging and selling as "St. Trinian's Finest Dry Gin". The headmistress then smells and tastes the product before asking for some bottles to be sent up to her office. As the headmistress leaves the lab, she comments that "practical things like chemistry provide such a natural outlet, I always think". This is followed by an explosion in the lab behind her, once again reinforcing the 'chemistry causes explosions' trope. Overall, *The Belles of St. Trinian's* (1954) is a disappointing depiction of chemistry and teachers in general; however, it may be the first depiction of 'breaking bad' in chemistry, *i.e.* using chemistry for illegal or underhanded purposes.

5.3   THE UNIVERSAL LANGUAGE

The idea that 'chemistry causes explosions' was not just prevalent in English-language movies at the time. *Caroline Cherie* (1951) is a French-language movie set in the late 18th century about a sixteen-year-old who falls in love with a young libertine called Gaston. In a brief scene at the beginning of the movie, we are shown a young man who is presumably Caroline's brother, experimenting in the attic of a large house. He explains that the famous scientist Lavoisier has promised a large sum of money to "the first chemist to produce pure hydrogen". Round-bottomed flasks, funnels and other equipment are clearly visible emitting thick condensate. We then cut away to a scene downstairs while the narrator explains that the risk of explosion was imminent in the attic, which duly occurs. The young man is later seen with a bandage around his head and nothing more is mentioned about this scene or chemistry for the rest of the movie.

However, it should be mentioned that not all onscreen chemistry from France during this era revolved around explosions. We also have another French-language movie called *L'Amour d'une femme* (1953), which opens in a school classroom where the local doctor is administering vaccines to the children. A new female doctor arrives on the island when the incumbent retires, with various scenes taking place in the doctor's office featuring cabinets filled with medicines and other containers. There are also scenes where the doctor is using syringes and other medical equipment, but there isn't really much chemistry to speak of.

Finally, we also have the Czechoslovak sci-fi adventure *Invention for Destruction* (1958) based on several works by the 19th century French author Jules Verne, later dubbed into English as *The Fabulous World of Jules Verne* (1961). The movie places real actors within animated/hand-drawn scenes and vehicles for a unique visual effect which is later repeated in a similar format in *Who Framed Roger Rabbit* (1988). The protagonist visits a professor at a sanatorium who explains that his role as a scientist means that his "interests lie in chemical reactions, not their practical applications". The issue of finances also appears once again when he declares that "I have no money left, but I must finish my invention". The scientist is kidnapped by pirates as well as the protagonist (an engineer) and during the chaos some beakers and a condenser are smashed.

They are both brought to a secret laboratory on an island where they are provided with everything they need to complete the invention. The lab is presented as a mix of drawings and real glassware, which is arranged for distillation with bubbling liquids and condensate present. The professor declares that "we search for the secret of matter itself... an experiment to release the enormous forces bound inside all matter". He is of course referring to atomic energy, but he plans for it "to serve mankind, to fuel machines, and provide light and warmth". However, he also warns that "how it will be used is for technicians and others to decide" while the camera focusses on an eagle with its wings spread sitting above a clock as a reference to Nazism. Later the professor draws a schematic of his invention which contains a large reaction vessel, condensers, and the correct circuit symbol for a battery power source.

Overall *Invention for Destruction* (1958) expertly references key issues relevant to the real-world scientific community, despite the fictional nature of the movie. In the real world, scientists were forced to work for the Nazis during WW2, who financed but also directed their research. This forced many to escape Nazi Germany, *e.g.* Einstein, Schrödinger, von Braun and others. In the movie, the first full-scale experiment fails and explodes, but a pirate exclaims "that's what I call results". Realising that the professor is unaware of the true intentions of the pirates, the protagonist creates a balloon using gas from a lab apparatus to carry a message to warn "world powers", who then "unite to combat the invention for destruction". Later, the professor sacrifices his own life to sabotage the new bomb made from his experiments, preventing the pirates from using it for war.

Finally, of note is a scene where a type of kinescope is used to project a movie for the pirates. While the projectionist is winding the handle, the projector catches fire in the background, but they continue talking unfazed. This is of course mocking the flammability of the cellulose nitrate film, which people had accepted by the time 'safety film' was finally introduced. Although 'safety film' (cellulose acetate) was first used for photography in 1909, it was not perfected for moving picture use until 1948, *i.e.* a decade before this film was released.

5.4   CHEMISTRY IN SCI-FI

After a brief absence, the early 1950s sees a return to stories by H. G. Wells with the first adaption of *The War of the Worlds* (1953) in full colour, based on the 1897 serial and 1898 novel of the same name. It opens with a narrator reviewing all the planets in the solar system that the Martians considered for migration, dismissing Neptune and Uranus due to their atmospheres of methane and ammonia, as well as Mercury because "the temperature at its equator is that of molten lead". The melting point of lead is 328 °C, while the surface temperature of the planet Mercury is now known to be 430 °C in the daytime and −180 °C at night. So, the figure of 328 °C is close to the real average of the highs and lows. The story begins when a meteor lands on Earth, which is discovered to be radioactive using a Geiger counter. Eventually the meteor turns out to be a Martian spacecraft which attacks using a "heat-ray", resulting in everything metal becoming magnetised.

The "machine from another planet" is powered by an "atomic source" and is later joined by a larger invasion which wreaks havoc on the Earth, resulting in a full-scale interplanetary war. However, there is a positive note mentioned throughout, that the

Martian invasion persuades all of humankind to work together for the first time to defeat a common enemy. This was a timely message at the time as the real world was still recovering from the aftermath of WW2. There are also numerous religious references throughout, similar to the original novel, which was heavily influenced by Darwin's publication *On the Origin of the Species* in 1859.

Interestingly, the movie made some changes to the source material, such as swapping the protagonist for a scientist instead of a layman. This speaks to where trust lay over fifty years after H. G. Wells published the original story, *i.e.* numerous scientific discoveries, such as DNA and fossils, had taken place which supported Darwin's theories. The scientist obtains a sample of Martian blood which he then examines under a microscope and claims that he might be able to use it to develop a virus. He and his colleagues also deconstruct a robotic Martian eye, which is demonstrated near some glassware such as funnels and condensers. The army also use the "atom bomb" against the Martians while accepting the danger of the radiation fallout. Although representing one of the first colour depictions of an atomic explosion in the movies, it is unsuccessful in destroying the Martians.

Following this, and referencing the real-life development of the atomic bomb as depicted in *Oppenheimer* (2023), the army turn to the scientists to find a solution. The scientists propose that they set up a remote laboratory in the Rocky Mountains (akin to Los Alamos) to "search for any weaknesses in the Martians". This results in a Christian religious reference when a scientist exclaims that the Earth will be destroyed in six days, to which the daughter of a pastor replies "the same number of days it took to create it". Before the scientists can begin, they are overcome by riots and chaos, implying that science could not do what God did. The movie ends with the Martians dying just before they destroy a church filled with people. The crowd in the church includes the scientists, who admit that they were also praying for a miracle. It is explained that the Martians succumbed to the germs in our atmosphere to which they had no immunity and "all that men could do had failed, the Martians were destroyed, and humanity was saved, by the littlest things, which, God in his wisdom, had put upon this Earth".

The concept of defeating alien visitors with Earth-based germs, viruses and/or bacteria that the aliens have no defence for, is an iconic idea by H. G. Wells. He was most probably inspired by the development of vaccines in the real world such as the smallpox vaccine by Edward Jenner (1796), the rabies vaccine by Louis Paster (1885), and the first cholera vaccine by Jaime Ferran y Clua (1885). This method for defeating aliens appears again later in other depictions such as *V: The Final Battle* (1984). It was also adapted for the computer age in *Independence Day* (1996), in which the protagonists upload a computer virus to the mothership which removes their protective shields.

Two years later, radioactivity and other atomic symbols also make numerous appearances in *This Island Earth* (1955), which is based on a trilogy of stories and a novel of the same name by Raymond F. Jones. The movie opens with the protagonist, an engineer (Rex Reason), explaining that he is working on the "combination of certain common elements into nuclear energy sources". His research laboratory mainly consists of electronic devices and materials, and we later learn that he is trying to turn lead into uranium. This a reference to the work of alchemists who were trying to turn lead into gold, implying that uranium was considered the 'new gold' at the time. There are also numerous references to the real-world technology boom of the 1950s, following the invention of the transistor in 1947, then the first modem in 1950, the first US nuclear plant in 1951, and the digital voltmeter in 1952, just to name a few. In terms of design, this period is usually referred to as the 'atomic age', which influenced architecture, industrial design, advertising, arts, and clothing.

In the movie, the researchers receive anonymous packages filled with advanced new technologies, as well as an instruction manual and blueprints telling them how to build an unknown device. A similar concept is later used again in *Contact* (1997) when the protagonist receives blueprints for a device hidden within radio signals. The device in *This Island Earth* (1955) is an elaborate video communication device which also allows the aliens to monitor and attack when necessary. Once built, it displays a symbol of the atom similar to the real-world 1950s US Atomic Energy Commission seal. The six light bulbs at the edge of each electron orbital implies that it represents a carbon atom, referencing the fact that humans are carbon-based life forms.

The interior of the autonomous airplane which carries the protagonist to meet the aliens is also strikingly similar to the one used in *Contact* (1997), where Ellie (Jodie Foster) also sits in a single chair with no seatbelt in an autonomous spacecraft that produces colourful lights while travelling. However, this is where the similarities between *This Island Earth* (1955) and *Contact* (1997) end. Nonetheless, it is very likely that Carl Sagan was inspired by *This Island Earth* (1955) when writing the story for *Contact* (1997), as he earned his Bachelor of Science in the same year (1955) before subsequently pursuing a Master's and PhD. The aliens in *This Island Earth* (1955) later claim to have recruited dozens of human scientists to help them "prevent future wars". However, the protagonist becomes suspicious of their true motives when he realises that they are only exploring nuclear energy. They then discover that the aliens are fighting a war with another species and need nuclear energy to maintain the protective "ionized layer" around their planet (Figure 5.1).

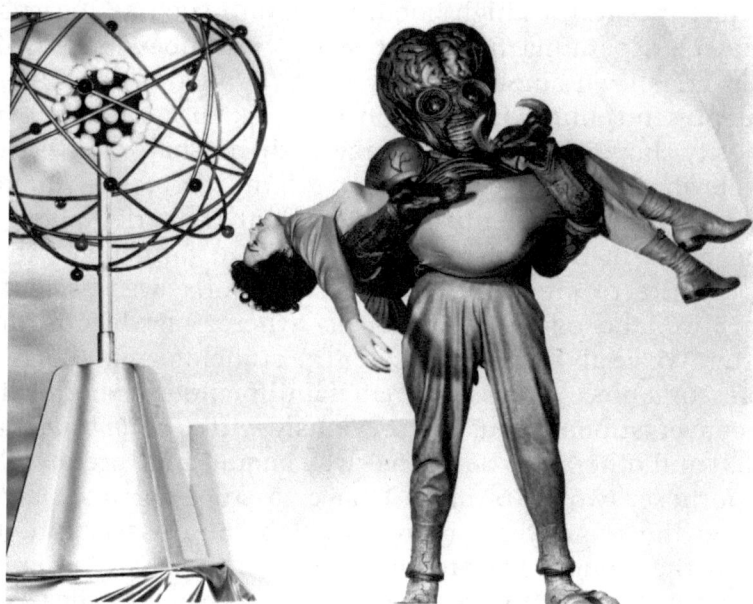

Figure 5.1   Images of the atom are frequently used in *This Island Earth* (1955). © Pictorial Press Ltd/Alamy Stock Photo.

This Island Earth (1955) also features a cat named Neutron, who is called that because "he is so positive". This is a strangely basic error since protons are positive and neutrons are neutral. Considering the level of detail provided throughout, this is more likely a poor attempt at a joke, since the cat is unfazed by everything going on around him, *i.e.* Neutron is a fitting name. There is also an interesting reference to neutrinos, which are described as "the missing link between energy and matter". In the real world, the neutrino was theorised since 1930, but the first experimental discovery was achieved by Reines and Cowan in 1956, a year after this movie was released. Reines and Cowan won the Nobel Prize in Physics in 1995 for this work and it is now known that neutrinos can indeed interact with an atomic nucleus to induce fission.

Next, we leave space behind to explore new worlds under our feet instead with *Journey to the Center of the Earth* (1959), the first screen adaptation of the 1864 Jules Verne novel of the same name. The movie is set in the late 19th century like the original story and opens in a geology lab at the University of Edinburgh where the protagonist (James Mason) is a geology professor who has just received a knighthood. Glassware such as measuring cylinders, separating funnels, beakers, test-tubes, and conical flasks are shown among various rocks on a lab bench. His student presents him with an unusual rock that he found in a curiosity shop, which appears heavier than usual.

Noteworthy is the continuation of the theme of financial difficulties among scientists. The student is worried about his appearance in front of faculty members at dinner due to his frayed shirt, to which the professor responds "we're scientists aren't we, the one part of society where frayed cuffs don't matter. We've all had them". Later the student complains to the professor's niece about his financial difficulties, echoing similar conversations discussed previously with *The Invisible Man* (1933) and others. He bemoans "why should I torture myself to no purpose, two years more I have to study for my Master's degree, then four years more as a laboratory assistant, then there's the money I have to give back to my relatives". He finds it difficult to see a future where he can provide for her as his love interest and therefore doesn't think she should be interested in him.

The professor runs numerous tests on the rock in his lab, with a large array of laboratory glassware shown onscreen filled with brightly coloured liquids. Some of the glassware also contain a condensate (again probably CO_2 for effect) while a red liquid bubbles through a tube alongside a green solution under a Bunsen burner with a yellow flame (lower temperature than a hotter blue flame). The professor explains that he has spent all day completing numerous tests on the volcanic rock, which prompts an interjection from his assistant: "Aye, and without the pause for lunch or tea". This again references the concept of scientific obsession and persistence, as seen previously. When asked for his theory, the professor proclaims that "scientists don't jump to conclusions, science is not a guessing game".

As he places the volcanic rock in the furnace, he orders his assistant to add "10 ccs of aqua regia". Aqua regia (Latin for 'royal water') is a real-world acidic, corrosive, and oxidative mixture of three parts concentrated hydrochloric acid (HCl) and one part concentrated nitric acid (HNO_3). It is commonly used by chemists and other scientists for cleaning and etching; mishandling can result in explosions due to its oxidising nature. Distracted by the conversation about the remaining time needed in the lab, his assistant adds all the aqua regia to the furnace by accident which results in an explosion. However, this reveals a message on a plumb bob inside the rock, which was left by a previous scientist who discovered a passage to the centre of the Earth through an extinct volcano in Iceland.

In one of the first depictions of scientific competition, the team race to Iceland fearing that a rival geologist is trying to steal their discovery. However, their rival is discovered poisoned with potassium cyanide by a descendant of the original scientist who discovered the route. Our protagonists identifies the potassium cyanide by its appearance and smell, which, to be fair, is quite distinctive since it smells like bitter almonds. Despite opportunities to discuss crystals and minerals, there are only brief chemistry references during the journey, one example being a deposit of cinnabar. This is a real-world mercury sulfide-based mineral which is usually found near volcanic rocks.

As they approach the centre of the Earth, they discover a "field of force which snatches all metal away", even non-magnetic metals like gold. All in all, they discover a vibrant and colourful world beneath our feet, including the "sunken city of Atlantis" where they find a giant asbestos-coated flame bowl, which they use to escape through an active volcano. Although damaging to human health, asbestos is indeed fire resistant and has been used for many millennia to coat ceramics. It was also mentioned previously in *Lost City of the Jungle* (1946) as a protective material for radiation. A few years later we have another geology themed movie called *Crack in the World* (1965), but this only contains brief scenes in a lab filled with glassware with little or no chemistry mentioned.

Finally, we have a brief mention of chemistry in the US sci-fi space movie *Forbidden Planet* (1956) when Bobby the Robot mentions that "Krell isotope 217" is lighter than Earth's lead, which is either an error or an intentional joke since an isotope with a molecular weight of 217 would be heavier than lead (207.2). However, *Forbidden Planet* (1956) is noteworthy for being the first depiction of faster-than-light travel, something which is central to the *Star Trek* franchise. It is also the first movie to be entirely set on another planet which orbits a star that is not our own sun. Finally, Bobby the Robot is also the first real depiction of artificial intelligence (AI) in a robot, and even has a distinct personality. Fictional elements, isotopes and other materials will be discussed again later.

5.5   MUTATIONS

Following the previous discussion about DDT in *Riders of the Whistling Pines* (1949), we have *Forever, Darling* (1956) about a research chemist named Larry, who is researching a new insecticide. In the sole lab scene, Larry is shown looking through a microscope in a lab surrounded by brightly coloured solutions in glassware. In the background there is also a periodic table from this era, *i.e.* with the correct number of elements for the year of the movie. Larry's superiors compliment him on a great job, calling his work a "crusade" and saying that "he hasn't been home for dinner in a month". This is another example of persistence and obsession, which are apparently required for success in chemistry. They also claim that his new insecticide will make "DDT look like talcum powder".

Later, Larry's wife tells some dinner guests that they are making do on a chemist's salary but they are not rich, despite the fact that they have a maid serving them dinner. Larry then declares that he has volunteered to run the field tests of the new insecticide for two years and wants to bring his wife with him. He tells their dinner guests "once you drop the ideal of money, all kinds of doors open, life becomes exciting, the view changes", continuing the ongoing reference to financial struggles in science as discussed.

The rest of the movie is mainly concerned with the relationship between Larry and his wife Susie, as they attempt to rekindle their marriage after Larry has neglected their relationship in favour of work. Later we learn that they intend to test the new insecticide on mosquitos that have developed resistance, claiming that "insect resistance is one of the toughest problems facing chemists in the world today... If insects could be controlled effectively, there could be an era of health and plenty such as the way the world has never known". There is also a brief mention of how a rubber dinghy inflates *via* the pull cord on a bottle of CO_2. Humorously, in the last argument before they make amends, Susie refers to Larry as a "Latin Louis Pasteur".

Away from chemical research, insecticides are also blamed for causing mutations. First, we have *The Incredible Shrinking Man* (1957) about a businessman who is enveloped with a bizarre cloud, causing him to progressively shrink in size. He visits a medical research institute where he drinks a "barium solution and stands behind a fluoroscopic screen". He is also given "radioactive iodine and an examination with a Geiger counter" as well as a "a paper chromatography test", all of which take place in a lab with glassware. The chromatography test reveals the presence of an insecticide which has "rearranged the molecular structure in his cells", causing him to shrink. It is concluded that a combination of insecticide and radiation from the bizarre cloud is the ultimate cause. The remainder of the movie is concerned with his survival as he continues to shrink, fighting off the house cat and various insects.

Along the same lines, we have *The Alligator People* (1959), which also features a lab filled with glassware. The movie opens in a medical institute where psychiatrists use sodium pentothal to help a nurse (Beverly Garland) recall repressed memories about her husband (Paul). The rest of the movie is then shown as a series of flashbacks about the nurse searching for her husband in a remote area of Louisiana. A home-based lab is prominent with shelves stacked with chemicals, as well as numerous round-bottomed flasks and other items connected by tubing like many previous examples.

We eventually discover that the doctor is trying to cure the alligator people of their disfiguration, which occurred when he injected them with reptilian hormones (hydrocortisone) and other chemicals extracted from alligators. He administrated the cocktail to them in order to regenerate their limbs after a plane crash. Hydrocortisone is a real-world pharmaceutical term for the hormone cortisol, which is used to treat severe allergic reactions by suppressing the immune system. Later, we see the glassware in the lab glowing, as well as various solutions bubbling and steaming over Bunsen burners with large flames.

Paul enters the lab and asks the scientist if the cobalt-60 has arrived. The proposed cure for the mutation involves a massive dose of radiation therapy by "combining X-rays with gamma radiation from the cobalt". Cobalt-60 is a real synthetic radioactive isotope of cobalt with a relatively long half-life of 5.27 years, produced artificially in nuclear reactors. In the real world it is used as a source of high-intensity gamma radiation for medical radiotherapy, as well as in the sterilisation of medical equipment and in industrial radiography. In the movie, the treatment goes terribly wrong, and Paul transforms further into an alligator.

Finally, no discussion about human mutation would be complete without a visit to Dr Jekyll, this time appearing as a series under the original name of the source material, called *The Strange Case of Dr Jekyll and Mr Hyde* (1955). However, there are only brief scenes in the lab, entirely in the first episode, with some close-up images of a mortar and pestle, and other scenes showing Jekyll transferring liquids between test-tubes and beakers. He also states that the concoction is "a solution of certain salts combined and balanced" and "the change in the solution's colour was not anticipated". Overall, this version marks the transition of the story further into the horror genre, with less emphasis on lab work. However, like previous versions, there are references to Jekyll's obsession and persistence, skipping meals and not emerging from the lab for two weeks. There are also other human mutation-themed movies from this era like *The Fly*

(1958), which was later remade with Jeff Goldblum in 1986, but there isn't really any other chemistry to speak of.

5.6   MONSTER CHEMISTRY

Famously, radiation is also the source of the mutation which produces the iconic Japanese monster in *Godzilla* (1954), re-edited as an American version called *Godzilla, King of the Monsters!* (1956). A Japanese sequel, *Godzilla Raids Again* (1955), also quickly followed. The discovery of strontium-90 in fish is stated as evidence for hydrogen (H-) bomb testing, which in turn is blamed for waking the prehistoric monster called Godzilla. Strontium-90 is indeed a product of nuclear fission, so it can be used as evidence of atomic bomb testing, but it does not specifically prove that fusion has taken place (*i.e.* in a H-bomb).

In the first *Godzilla* movie, numerous scenes take place in a basement lab where a scientist is first shown grinding a substance in a mortar and pestle near condensers and other glassware, along with some solutions bubbling. Other scenes show more glassware with tubing, shelves of chemicals, and animal test subjects in the form of fish and a lizard. It is later explained that the scientist has discovered a "terrible chemical" which rids water of oxygen, destroying all aquatic life. He attempts to keep it secret, fearing the consequences of such a discovery, most likely referencing those who invented the atomic bomb as seen in *Oppenheimer* (2023). He eventually relinquishes the secret in order to defeat Godzilla and warns that it will only be used once, as he destroys all his notes.

A few years later we also have a British version of the *Godzilla* story in *The Great Behemoth* (1959), which opens with a nuclear explosion and a scientist (marine biologist) explaining the dangers of the atomic age. He explains that "one millionth part of a gram of radium is the safety limit, the amount that one human can tolerate". He warns that the nuclear waste and the radiation fallout from testing is having a substantial effect on marine life, which he calls a "biological chain reaction". Later, we are shown a scientific lab with glassware, tubes and bubbling solutions where they are testing seawater and fish samples. The method used to analyse the sample for radiation is explained as "we evaporate the liquid and examine the residue. This new ionisation chamber is very sensitive and shows the slightest deviation from normal radiation". Eventually it is discovered that the radiation fallout has woken an ancient dinosaur that was lying dormant for centuries, which wreaks havoc on mankind.

Monster themed movies remained popular throughout the 1950s. We even have Clint Eastwood playing the role of a lab assistant in *Revenge of the Creature* (1955). In his first-ever acting role, he blames a cat for eating a rat that he later discovers in his coat pocket. He only appears in a single scene wearing a lab coat against a backdrop of lab glassware and large containers. Towards the end of the movie, the scientists are shown working in a lab with test-tubes as they compare blood samples, but that is the extent of the onscreen chemistry.

In the same year we also have *Tarantula* (1955), which features numerous scenes in a lab where a scientist is attempting to increase the size of various animals using nutrients. Interestingly, a glove box is shown where the scientist is extracting a substance from a beaker using a syringe, which he then injects into a monkey. Test-tubes, sample containers and conical flasks are shown around the lab as well as various animal test subjects which appear much larger than usual. The scientist is attacked by his colleague, which accidentally releases the giant tarantula, setting up the theme of the movie.

The scientist claims that his research was aimed at feeding the "overcrowded world" through the development of a new nutrient and professes that no-one can do it on their own: "you build on what hundreds of others have done before you". He also explains that he uses a radioactive isotope that he calls "ammoniac" to help synthesise the nutrient. Later, the professor and his new assistant are seen wearing lead aprons with face shields as they place the radioisotope in the glove box to collect it in a syringe as before. It transpires that the professor's colleagues already tested the unstable nutrient on themselves which resulted in their deaths. While investigating the giant tarantula, the protagonist travels to an agricultural lab where more glassware is set up in the background, once again connected with seemingly random tubes. Eventually, the tarantula is defeated with an airstrike.

Finally, two years later we have two more monster movies with some onscreen chemistry called *The Black Scorpion* (1957) and *The Monster that Challenged the World* (1957). The first is about giant scorpions who escape from a volcano and wreak havoc. It contains a few tokenistic scenes in a lab filled with glassware involving a Mexican scientist, who tells the protagonist about the poison they found. Once again, the glassware is connected together with random tubes, *e.g.* the outlet of a Buchner flask is connected to the top of a condenser, all suspended high in the air on a retort stand. He says the poison is not "chemical", but "organic". The scientist then tells his assistant to fill a test-tube with pure alcohol, another with distilled water, another with tequila, and one with salty water. When asked about the tequila, he says "In your country, I believe you call it a coffee break".

In *The Monster that Challenged the World* (1957), the monster takes the form of giant water mollusk, which is once again the result of atomic testing (and an earthquake). The onscreen chemistry takes place in a US Navy laboratory on a remote base which features test-tubes and conical flasks (again linked with tubing). There are also numerous separation funnels held in place with retort stands. When accused of making the water radioactive from their atomic testing, the scientist defends their work by proclaiming that "science fact and science fiction is not the same". Later, the scientists recover an egg from the creature, which accidentally hatches, devours all the animals, and destroys the glassware.

Finally, we return to the Swiss mountains for an old gothic classic in *The Curse of Frankenstein* (1957), which explores the origins of the famous scientist. The young Baron Victor Frankenstein hires a tutor and the first lesson is chemistry. On the chalkboard we can see several benzene chemical structures and the incorrect chemical equations for the reaction of copper metal with sulfuric acid as $Cu + H_2SO_4 \rightarrow H_2 + CuSO_4$. The real products should not include hydrogen gas (H_2), and instead should be water and sulfur dioxide as well as copper sulfate.

This then leads to the older Victor Frankenstein (Peter Cushing) in an attic lab with his tutor. We can see several funnels, Büchner flasks, conical flasks and beakers containing a red substance. The red liquid is also seen bubbling through the apparatus, and they use electricity to cause a bath of steaming liquid to bubble. We then discover that they were attempting to revive a dog inside the bath. After their success, the tutor is excited by the results and starts postulating how this

can be used in surgery. However, Victor wants a bigger challenge, to revive multiple body parts to create a new creature.
The tutor initially goes along with the scheme but proclaims
reservations and then refuses to continue.

Eventually, Victor's obsession with the experiment drives him
to commit murder to obtain the brain of a famous scientist for
his creature. Interestingly, there are no lab coats, safety goggles
or wild white hair portrayed in this version of the story. Victor is
called "a mad man" by his tutor though, in keeping with the
concept of 'mad scientist'. Later, we also see more glassware
filled with colourful liquids, in addition to condensers and
chemical containers. We also see colourful powders in containers, along with tubing joining some of the glassware together. This use of glassware will become a prominent feature as
time goes on.

SEASON 6

Colourful Chemistry

6.1 LIFT-OFF FOR SCIENCE FICTION

As the 1950s ended, and the number of accurate depictions of onscreen labs and glassware decreased, the golden age of onscreen chemistry was coming to an end, making way for the sci-fi era. As sci-fi movies started to re-imagine spaceflight and exotic far-off worlds, onscreen chemistry continued to be influenced by real-world events, more so than other sciences. The golden era of movies made way for the golden era of science fiction (sci-fi), with the release of box offices hits such as *The Time Machine* (1960) by H. G. Wells and *Planet of the Apes* (1968).

The influence of the atomic age also made way for the space age, spurred on by the real-world space race between the US and Russia with the launch of *Sputnik 1* in 1957, Yuri Gagarin in 1961 and *Apollo 11* landing on the moon in 1969. The new public fascination with space influenced movies and TV, giving us *Star Trek* (1966–1969) on TV and *First Men in the Moon* (1964), another H. G. Wells story, in the theatres. Movie makers also broke new ground with accurate depictions of space flight in *2001: A Space Odyssey* (1968).

The first colour animation on TV also appeared with *The Jetsons* (1962–1963, 1985–1987), about a family living in Orbit City 100 years in the future. The Jetsons captured the technological

Onscreen Chemistry: The Portrayal of Chemical Science in Film and TV
By John O'Donoghue
© John O'Donoghue 2025
Published by the Royal Society of Chemistry, www.rsc.org

optimism of the atomic age and coupled it with the space age. It dreamed about a future filled with gadgets and invited audiences to imagine the dazzling possibilities that technological innovation could achieve. Even the names for everyday objects reflected the era of innovation, like "cosmic cola" and "molecular motors". Noteworthy also is that the Jetsons had a nuclear-powered robotic dog called Astro.

Although we don't have nuclear-powered robotic dogs today, they did predict robotic vacuum cleaners, mobile phones, self-driving cars, and video calls. Interestingly, the video calls featured a mask that would cover your 'morning face', akin to the filters now common across social media platforms. There were also examples of ordering food from touchscreens, as well as "dial-a-meal" which produced food in a pill form. The depictions of pills and other drugs, both medicinal and recreational also begins during this era, shifting the image of chemistry into a new realm for the first time.

6.2   SPACE MONSTERS

Although the 1960s moved away from monster movies, there were still a few concepts left to explore. The first is *Dogora* (1964), from Japan, about an "amorphous cellular life-form" which consumes carbon in the form of coal and diamonds. The creature occurs due to "a pocket of radioactive matter in the atmosphere", where "a certain kind of cell exists which has reacted abnormally to the presence of cobalt and strontium to create this energy". It is explained by a scientist that coal and diamonds are similar in that they are both crystalline, but if the lifeform goes after all carbon-based material, then the world is in great danger.

Later, in a lab, there is plenty of glassware on benches (no tubing!) including test-tubes in racks, but, unusually, they are all empty. Later, when the army are unsuccessful in their attack, they turn to the scientists to find a way of stopping the creature. We return to the lab where everything is now filled with colourful solutions and there are numerous scientists in lab coats looking through microscopes. They observe the space cells transforming into crystals, which occurred when "a large swarm of wasps attacked the space cells and destroyed them. The wasp toxin must have caused some chemical changes in the space cells". This leads to a lab montage with scientists in full environmental suits working with lots of colourful solutions in glassware and a close shot of two round-bottomed flasks in a water bath. It is explained that the toxin was isolated and sent for full scale production, which eventually defeats the lifeform, causing it to rain colourful rocks as the cells are transformed into crystals.

Next, we have *Evil Brain from Outer Space* (1964), which is an edited compilation of films seven, eight and nine of the very popular Japanese *Super Giant* (1957–1959) film series. The protagonist for these movies is the Japanese equivalent of Superman called Super Giant (called Starman in the US and Spaceman in Europe) and the three original movies which comprise the edit are entitled *The Space Mutant Appears* (1958), *The Devil's Incarnation* (1959) and *Kingdom of the Poison Moth* (1959). The first six movies, on the other hand, concern topics such as terrorists with atomic bombs, reptile invaders from space, changing the Earth's rotation and spy satellites. These were also edited into three films and dubbed into English for US audiences. But it is the final three movies comprising the US compilation *Evil Brain from Outer Space* (1964) which has some onscreen chemistry.

The three plots are woven together into a story about a marauding alien brain-like creature and his mutant army, helped by evil scientists. The evil scientists are only shown in secret labs (usually in a basement), working with some glassware containing bubbling solutions, condensate, and tubing. Also, they are almost always wearing lab coats and one has an eagle perched on his shoulder accompanied by a one-legged assistant, completing the image of pirates. But we also have good scientists shown working with glassware like test-tubes, condensers, volumetric flasks and beakers in a large lab with a chalkboard and numerous instruments in the background. The good scientists also appear outside the lab and remove their lab coats on occasion. The voiceover explains that "scientists continue desperately to seek one formula to prevent the mutants from dividing and another formula which will destroy the brain". The mutants are described as having "solid cobalt nails" which Super Giant needs to avoid. Eventually, with the help of Super Giant, the good scientists pour the formula over the brain which dissolves in a cloud of smoke.

Next, we have the British monster movie *Quatermass and the Pit* (1967), based on the TV serial of the same name. It revolves around the discovery of unusual skulls in the London Underground alongside a Martian spaceship. While carrying out some tests on one of the skulls in a lab, we can see some glassware attached to retort stands and joined together with tubing. Later, when they are dissecting the Martian creature, beakers, conical flasks, condensers, funnels, and test-tubes are shown filled with coloured liquids. Interestingly, when discussing why the creatures landed on Earth, our protagonist asks "If we found that our Earth was doomed, say by climatic changes, what would we do about it?", to which the other character responds "Nothing, just go on squabbling as usual".

Finally, from the original *Star Trek* (1966–1969) TV series we have *The Devil in the Dark* (season 1, episode 20), which features a "silicon-based lifeform". When Spock analyses a fragment from the creature, the composition is claimed to be like "fibrous asbestos". For Trekkies, this episode is also significant because it marks the first appearance of Doctor McCoy's famous catchphrase, "I'm a doctor, not a...", in this case "a bricklayer", when he is asked to heal the creature with a silicon-based cement. Although silicon is in the same group of the periodic table as carbon, its bulk prevents it from forming the same molecular structures that carbon-based life uses. Silicon is also inert at room temperature, but in this episode the creature resembles molten rock, lending some credibility to the concept.

6.3   MORE CHEMISTRY INVENTIONS

We return to the theme of chemistry inventions with *Lover Come Back* (1961), but this romantic comedy has little in common with the examples discussed previously, *e.g. The Man in the White Suit* (1951). The protagonist is Jerry (Rock Hudson), a marketing executive, who comes up with a non-existent product called "VIP" to beat a rival marketing agency. However, when the non-existent product accidentally gathers interest, he turns to a fictional and eccentric "chemistry Nobel Prize winner" called Dr Linus Tyler (Jack Kruschen) to invent something. Dr Tyler is described as "a troublemaker, a dangerous non-conformist, he has been fired by three companies. Money can't buy him, he's incorruptible" – quite the compliment!

Our chemist is mainly shown working in the lab filled with glassware like volumetric flasks, beakers, conical flasks, separation funnels and more. Of note, there are conical flasks filled with coloured solutions sitting on tripods and attached to retort stands high off the bench, presumably to keep them within the frame of the camera. Dr Tyler is feeding his lab hamster with something from a beaker, which he also eats himself, as Jerry places money on the bench. Dr Tyler claims that he prefers to remain away from the spotlight, stating that he has "quit the world – they didn't appreciate me when they had me, now let them suffer".

Dr Tyler bemoans the demands of marketing and how it conflicts with scientific integrity by telling Jerry that he once "invented a hair tonic superior to anything else on the market, but would they buy it? Not until they were told it contained a secret ingredient". So he added something called TR2748: "Do you know what TR2748 was? My phone number!". As Jerry tries to convince him to invent VIP, he claims that he will never again "prostitute my genius. Not for all the gold on Madison Avenue". However, he quickly concedes when he is offered the entire envelope of money ($5000), once again a reference to the financial troubles of scientists and implying that scientists are easily bought.

A chalkboard with semi-nonsensical chemical equations is briefly shown on screen. Although CH_3I is a real compound called iodomethane (an insecticide), it is shown reacted with M_6, but there is no chemical element with the symbol M; the most similar is molybdenum (Mo). The product (CH_3M_6I) also doesn't make much sense even if M represented 'metal'. It is then reacted with carbon dioxide (CO_2) to form acetate (CH_3CO_2) and M_6I. The equations are therefore a mix of real and fictional chemistry. Jerry is mistaken for Dr Tyler by a woman from the rival agency, so he plays along, pretending to be the chemist. Jerry convinces her to meet elsewhere to discuss VIP because the lab is too dangerous, to which she responds "You mean there could be an explosion?", and Jerry replies "Oh yes, if I'm not careful, this whole thing could blow up".

Later, when we revisit our eccentric chemist in the lab, there is an explosion from the basement window (where the lab is located) onto the street. As Dr Tyler attempts to figure out his mistake, the almost correct structure of 1,3,5-trimethylbenzene is shown on the chalk board, although he is missing a few hydrogen atoms. He then rubs off the 3 from one of the methyl groups and changes it to a 4, making it chemically impossible (Figure 6.1). After several mishaps, Dr. Tyler is described as "the hermit", "the mad chemist" and "your Frankenstein friend" by others, in addition to "he's not a chemist, he's a frustrated munitions maker". Numerous colourful explosions later, Dr Tyler invents an alcoholic mint sweet as strong as a triple martini, setting up the antics for our screwball comedy. He delivers the product just in time for a meeting with the ad agency, the only time that he appears outside of the lab.

Overall, much of the main theme of the movie is repeated in *What Women Want* (2000), starring Mel Gibson and Helen Hunt, but without a chemist or a lab. In contrast to other chemistry invention-themed movies, Dr Tyler is depicted as easily bought, mad, and eccentric, and overall *Lover Come Back* (1961) centres around the 'chemistry causes explosions' trope.

In the same year we also have *The Absent-Minded Professor* (1961), later remade as *Flubber* (1997) starring Robin Williams. The original movie opens in a university lecture theatre with Professor Brainard carrying out a chemistry demonstration with glassware, lots of smoke and tubing. In the background a model of a crystal lattice structure (*e.g.* sodium chloride) is briefly seen as the professor explains different forms of energy. The demonstration explodes and the professor continues unfazed.

Described as a "physical chemist", Brainard also has an elaborate glassware setup in his home shed, with glowing

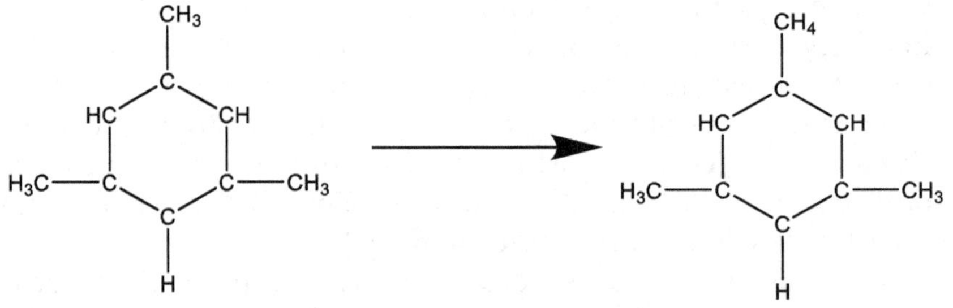

Figure 6.1   The molecular structures shown on the chalkboard in *Lover Come Back* (1961), where the chemist changes the number of hydrogens on one of the methyl groups.

solutions, flashing lights and strange sounds, strikingly similar to *The Man in the White Suit* (1951). The glassware is also joined together with random tubing, like many previous examples. The chalkboard contains several organic molecules, mixed among random equations, making it difficult to decipher anything specific. The obsession and persistence trope also makes another appearance (now generally only seen in male chemists), causing him to miss his own wedding for the third time as his home lab explodes.

There are also numerous unconnected (and sometimes nonsensical) references to scientific terminology throughout the movie, *e.g.* "love is a bouquet of lovely hyacinths, not a beaker of heavy hydrogen". The explosion in the shed results in the discovery of a "a new kind of energy" (later described as repulsive energy) in the form of a flying rubbery material (flubber), setting up the antics for the movie. Brainard describes it as "the application of an external force triggers a molecular change, liberating energy, a type previously unknown". One of the tests he conducts on the flubber is bombarding it with gamma rays from an "ordinary radioactive isotope" (in his shed). Interestingly, the financial difficulties this time are with the college, not the professor, with flubber coming to the rescue as a valuable invention.

One of the central applications for flubber throughout the movie is a flying Model T car, a feat which is repeated some years later in the well-known *Chitty Chitty Bang Bang* (1968). Brainard also attaches it to the heels of the shoes used by the college basketball players, which helps them jump much higher, the first chemistry and sport cross over since *It Happens Every Spring* (1949). It is also worth noting that Brainard does eventually go through with his wedding, once his invention has been recognised by the college president and the college is saved from an unscrupulous businessman trying to use flubber for personal gains.

Next, we have *Caprice* (1967) about industrial espionage involving a secret cosmetics formula, double agents, and rival companies. The movie opens with our protagonist (Doris Day) getting caught trying to sell the formula for a new roll-on underarm deodorant. She is fired from a European cosmetic firm and picked up by a US firm instead. Later, she is drugged by the new US cosmetics firm using a syringe and an unspecified sample vial. While under the influence of the unspecified drug, she is asked questions about her former employer. Later we meet the French inventor of "Caprice" who runs a lab behind her shop.

The lab is presented with numerous perfume bottles on shelves, and some large round-bottomed flasks clamped in place near a mortar and pestle, all filled with boiling green liquids, but no tubing. Lilia Skala is working on "a new formula for eye shadow" as she talks about her family history of inventing various cosmetics. She proceeds to add a liquid into a measuring cylinder while holding it correctly at eye level to prevent parallax. She also explains that she developed the water-repellent hairspray Caprice for herself because she does a lot of skiing and didn't want her hair getting wet. This is the only real onscreen chemistry scene in the movie, but it is a commendable depiction. Finally, we discover that the US cosmetics firm was hiding drugs in their face powder, which, when burned and reduced to ashes, releases "a powerful hallucinogenic, distributed all over the world in an innocent cosmetics box".

In the same year, we also have *Clambake* (1967) about an heir to an oil fortune (Scott) played by Elvis Presley, who trades places with a water-skiing instructor (Tom) to see if the girls will like him for himself rather than for his money. He decides to fix a damaged boat to enter a boat race, explaining to the owner that the resin on the hull was the problem and introduces him to "GOOP". He goes on to say that he "majored in engineering and got a job in a research lab after I graduated" so he wants to "try some experiments" in the owner's lab. This leads to a lab montage complete with glassware, coloured solutions, condensate, Bunsen burners and lots of tubing. Elvis even wears the lab coat with the collar turned up! There is a brief pause in the montage when Tom attempts to take a drink from a beaker, but when Scott attempts to stops him, Tom tells him it's coffee.

Later we discover that his father wants to sell "all completed and incomplete experiments, formula and data on our ill-fated hunt for a glycol oxyoctanoic phosphate (GOOP)". We return to Scott who is asleep in the lab with a centrifuge spinning next to him, surrounded by glassware. A package arrives from his father's research lab with two bottles of his previous experiments which allows him to finish the work and coat the boat. The GOOP needs 24 hours to harden, so there is no time for an experiment, but he declares that he is confident in his work. It works and they now plan to market GOOP. In the end, Scott comes clean to his love interest, representing a very different financial conversation related to science and research compared to all previous depictions. Scott's invention is not for financial reward or recognition, he is genuinely interested in it and helping others.

In the real world, the slang name for boat resin is in fact 'goop', owing to the gloopy viscosity of resin, but there are also numerous commercial products also using the same name. We can't be sure of what the molecular structure of glycol oxyoctanoic phosphate would be in the movie, but all three words are real, so a variety of different structures are possible. Finally, although not necessarily an invention, worth a brief note here is the homemade distillation apparatus in *The Great Escape* (1963), humorously used to make moonshine from the peels of potatoes. Distillation is of course a common chemistry experiment for purifying or extracting.

6.4 SELF-EXPERIMENTATION

As sure as death and taxes, we kick off the 1960s with another version of the now well-trodden story of Dr Jekyll. The first colourised version of the story is a British horror called *The Two Faces of Dr. Jekyll* (1960). Like the 1940s version, the garage-based lab also contains live animal test subjects, but significantly less glassware this time. Like previous screen adaptations, there are numerous volumetric flasks with coloured liquids on shelves as well as a conical flask filled with a dark liquid bubbling over a Bunsen burner with a yellow flame. The flask is attached to a condenser which is distilling a colourless liquid off screen to another tube. Overall, the glassware setup is more sensible compared to previous depictions, and the active chemistry is realistic. There is also a scene later in the movie where Dr Jekyll talks to Mr Hyde in the mirror, which is another possible reference to molecular chirality, as discussed previously. The chemical concoction also ages Jekyll rapidly, with his hair getting progressively greyer as the movie continues.

There are similarities to previous depictions, such as Dr Jekyll skipping dinner parties and spending time alone in his lab. However, there are notable and important differences in terms of how scientists are depicted on screen. Firstly, his wife (Kitty) is depicted as cruel and unsupportive, carrying on a secret relationship with his friend. Dr Jekyll asks her "Let's both take the evening off, you from being social and me from being antisocial. Let's be together tonight", but she continues on to the party and responds "You may not need friends, but I do". His obsession with work and neglect of his wife is only briefly mentioned. This version also changes the method of administration of the chemical concoction from a drink to a syringe. Also, Mr Hyde is not grotesque in this version, instead he is dashing and suave, but retains his psychotic and unpredictable aggression. Finally, there are numerous references made to "opium addiction", which rose to prominence in the real world during the 1960s.

In the same decade we also have *The Strange Case of Dr. Jekyll and Mr. Hyde* (1968), a Canadian–US television film. Early in the movie, Jekyll is presented in his basement lab watching a large condenser. As the camera pans around we also see large round-bottomed flasks filled with red, green and purple liquids, connected with random tubing. Various solutions are also bubbling over Bunsen burners and the shelves are stacked with chemical bottles. After Jekyll pours a suspension into a conical flask, he takes some notes in his lab book. He then proceeds to pour it into a beaker before drinking it back. Later, he attaches a funnel with an airlock to a flask, which he then uses to add some blue liquid while stating "this new distillation will be ready by morning". However, there are also measuring cylinders containing bubbling liquids, which makes no sense. The rest of the movie is largely similar to the 1930s' and 1940s' versions of the story.

Continuing the theme of self-experimentation, we have a modern re-interpretation of the Dr Jekyll and Mr Hyde story in *The Nutty Professor* (1963). It opens in a similar fashion to *The Absent-Minded Professor* (1961), *i.e.* with a professor doing a chemistry demonstration lecture. Lots of colourful solutions in various forms of glassware are shown, once again joined together with nonsensical tubing. Inevitably the entire experiment explodes violently, resulting in the fire service breaking down the door to rescue the students. However, unlike nearly every previous depiction of chemists onscreen, our protagonist (Prof. Kelp) is presented here by Jerry Lewis as a socially awkward, bumbling, hunched, buck-toothed and squeaky voiced nerd (Figure 6.2). This depiction may well have been the inspiration for Professor Frink in the animated TV and movie series *The Simpsons* (1989–present), which will be discussed later.

One of his experiments is described as "developing an additive to gasoline, for the purpose of increasing the efficiency of the combustion engine". The formula of the additive is given verbally as $C_3H_5NO_3$ and described as nitroglycerin, which is incorrect – the real-world formula for nitroglycerin is in fact $C_3H_5N_3O_9$. Later, we have a scene in the chemistry teaching lab which contains plenty of red and blue condensers, as well as green solutions. When refused permission to leave the class for football practice, the jock picks up Prof. Kelp and places him in the solvent cabinet. In a follow up scene Prof. Kelp describes

Figure 6.2 Jerry Lewis playing the role of the *The Nutty Professor* (1963) with an elaborate glassware apparatus with lots of tubing. © Pictorial Press Ltd/Alamy Stock Photo.

hydrogen as a powerful explosive, as well as nitrogen, which is incorrect since nitrogen is an inert gas (non-reactive). However, behind him on the chalkboard there are correct half-reactions demonstrating the oxidative states of nitrogen, so he may be referring to these.

After failing to build any muscle in the gym, Prof. Kelp then turns to "chemistry and pharmaceuticals" instead, because "that's what I know". He consults numerous medical textbooks and comes up with a formula which he begins to produce in his university lab. In a homage to Dr Jekyll's transformation into Mr Hyde, Prof. Kelp is shown working alone in the lab, surrounded by measuring cylinders, test-tubes and tubing filled with coloured solutions. Before drinking his formula, the pet bird in the lab warns him to "think about it", a superficial version of the warning given in *Monkey Business* (1952). He proceeds to drinks the pink liquid from a beaker, causing him to transform in a similar fashion as Dr Jekyll into Mr Hyde.

Like the dashing, suave and confident version of Mr Hyde portrayed in *The Two Faces of Dr. Jekyll* (1960), Prof. Kelp's alternate personality is also suave, confident, aggressive and brash, calling himself Buddy Love. He begins to pursue one of his female students (Stella), who previously showed some interest in Prof. Kelp. Like most versions of Dr Jekyll, Prof. Kelp knows that Buddy Love is a bad person but cannot stop himself from taking the formula since he enjoys the attention it gives him, and he thinks Stella is only attracted to Buddy. The formula wears off at inopportune moments and eventually his alternative personality is exposed. Interestingly, we also discover that Stella prefers Prof. Kelp over Buddy Love. Finally, Prof. Kelp's father attempts to market the formula, while Kelp and Stella slip away with the formula to get married.

Together, *The Absent-Minded Professor* (1961) and *The Nutty Professor* (1963), although fun and light-hearted, contribute significantly to the association of chemistry with the negative image of 'mad scientists' and 'nerds'. In many ways, they are almost like a collection of the worst parts of previous onscreen chemistry depictions, reinforcing many false stereotypes. Both movies also continue the link between chemistry and explosions, with little in the way of real-world chemistry or lab skills. Even the 1950s mutant movie *The Alligator People* (1959) correctly explained which radioactive isotope they used as a source of gamma rays, *i.e.* cobalt-60, unlike *The Absent-Minded Professor* (1961).

Worth a brief note here is the sequel to *The Absent-Minded Professor* (1961) called *Son of Flubber* (1963). This time the professor uses "flubbergas" to influence the weather. This was quite likely inspired by cloud seeding in the real world, which gained significant traction in the late 1950s.[1] This time the sport crossover is American football, but there's even less onscreen chemistry than last time. Interestingly, the professor and the college still struggle financially since they haven't received any money from the invention of flubber.

6.5 BOND, CHEMICAL BOND

The 1960s also gives us the first onscreen appearances of the most famous spy in the world, James Bond. Fitting for a person named 'Bond', there are numerous examples of chemistry and labs throughout the entire series of movies, kicking off with the first movie *Dr. No* (1962) starring Sean Connery as Bond. Our first reference is when an enemy agent kills himself by biting into a cyanide-laced cigarette before Bond can interrogate him. Later, Professor Dent is introduced as a "metallurgist" who runs a "test laboratory", which is briefly shown on screen containing a weighing balance and some chemical containers. The dishonest Dent explains that he was asked to test some rock samples, which he claims were just "low-grade iron pyrite", better known as fool's gold in the real world.

Using a Geiger counter, Bond discovers that the samples are radioactive. Under the cover of darkness, Bond visits the island where the rock samples originated and meets a seashell diver, Honey Ryder (Ursula Andress). Both are captured and taken to a hidden base, where they are put through radioactive decontamination and tested numerous times with a Geiger counter. They are then knocked unconscious with "drugged" coffee (a common feature of Bond movies). They then meet the base's owner, Dr Julius No (Joseph Wiseman), who is a Chinese-German criminal scientist with prosthetic metal hands due to radiation exposure. We are then provided with an interesting physics reference when the glass of the aquarium is described as "convex,

10 inches thick, which accounts for the magnifying effect". Bond describes this as minnows pretending to be whales, "like you and this island", but Dr No corrects him: "It depends, Mr Bond, on which side of the glass you are".

Dr No explains that he offered his services to both the 'East' and the 'West', but they refused and now he wants to take revenge. Bond escapes detention and uses a decontamination suit to infiltrate the control room where we see people using telescopic control arms with radioactive substances in a sealed room. Bond stops the missile launch by pushing the control rods entirely into the nuclear reactor, producing large amounts of steam and destroying the base in an explosion.

The next Bond movie in the series is *From Russia with Love* (1963), but other than more drugged food items, poison-tipped blades and a brief mention of a tear-gas cartridge by Q, there isn't much in the way of chemistry. The most notable is the use of chloral hydrate as a sedative, which is described as "quick but mild". However, in the real world it takes between 20 and 60 minutes to be effective. The next onscreen chemistry we find is in *Goldfinger* (1964), with a brief scene inside Goldfinger's metallurgy workshop where some empty laboratory glassware can be seen on a bench in the background. It seems to be part of a lab, with instrumentation on the other benches against the wall along with extraction hoods. The glassware apparatus looks like it is set up for distillation.

In the original novel, the titular villain plans to spike the water supply of Fort Knox with "GB", which is described as "the most powerful of the Trilone group of nerve poisons". However, in a striking throwback to *Things to Come* (1936), the screen adaption of Goldfinger releases a gas from an airplane to render the staff of Fort Knox unconscious. The gas is called "delta-9 nerve gas", which Goldfinger initially claims only induces unconsciousness for 24 hours, but it later transpires that it will kill everyone. He first uses it on a small scale in his boardroom to dispatch the gangsters who sold him the nerve gas, which prompts Bond and Pussy Galore to switch the canisters at the last minute before it is released over Fort Knox.

It transpires that Goldfinger has no intention of stealing the gold from Fort Knox; instead, he wants to detonate an atomic bomb inside to render all the gold worthless due to radioactivity. This would then increase the value of his own gold. In the final show-down, the inside of the atomic bomb is shown as a collection of mechanical wheels, spinning disks and flashing lights. Bond attempts to pull apart some cables to stop it, but a CIA agent comes to the rescue and switches it off with 007 seconds left on the clock.

Nerve agents also make an appearance in the next Bond title, *Thunderball* (1965), which introduces a fictional nerve agent called "Gamma Gas", described as "instantaneous". This is another switch from the original novel which specifies cyanide gas. The Gamma Gas is used twice during the film: a spray gun is used to kill a NATO Major, so he can be impersonated aboard a Vulcan bomber to steal two atomic bombs. Once airborne, a cylinder of the Gamma Gas is introduced into the aircraft's air supply, flowing through the other occupants' oxygen masks, killing them instantly.

After that we have *You Only Live Twice* (1967), which only shares the name with the original novel since most of the story was changed for the screen. The space age really comes to the fore here, when Bond is sent to investigate the mystery of a rocket which is capturing US and Russian orbiters. In Japan, Bond sneaks into the Osato Chemical Engineering Co. Ltd and discovers documents with an order for "LOX". Bond correctly explains that 'lox' is an American term for smoked salmon, but "it is also the technical name for liquid oxygen, which makes rocket fuel". He returns undercover as a businessman from Empire Chemicals looking to produce the food additives "monosodium glutamate and ascorbic acid".

Later, at the docks there are large metal tanks shown with the words "Osato Chemical Co. Synthetic Turpentine" stamped on them, but Bond spots condensation on the tanks, exclaiming "ice cold, liquid oxygen". The last chemistry reference takes place when Bond and the Japanese agent attempt to access the secret base *via* a tunnel. However, before they enter the tunnel Bond smells "phosgene gas", a real-world insecticide which was also used during WW1 as a chemical weapon. In low concentrations it smells musty, resembling freshly cut hay or grass. The rest of the movie is then concerned with stopping Blofeld from trying to start a war between the US and Russia.

The 1960s Bond movies finish with *On Her Majesty's Secret Service* (1969), with George Lazenby taking over the duties from Sean Connery. This time our villain, Blofeld, has established a clinical allergy research institute in the Swiss Alps which claims to cure various allergies. In a brief scene, Bond is brought down into the mountain to meet Blofeld and they pass through a purple light (presumably UV) which is explained as "antisepsis". During their conversation, a lab filled with glassware and scientists in lab coats can be seen in the background through the window. Lots of conical flasks filled with a red solution are on a large agitator instrument used for mixing. Blofeld explains that he is producing individualised vaccines for allergies. However, it later transpires that he has developed an infertility virus for biological warfare which stops reproduction in plants and animals. We briefly see the lab in more detail later as Bond fights his way through it, showing some flasks clamped to retort stands. But that's all the chemical Bonds for now, we will return to Bond again in the next chapter.

Worth a brief mention here also is a future Bond, Roger Moore, who plays the lead role in *The Saint* (1962–1969) TV series, based on the character Simon Templar first created by Leslie Charteris in the 1920s. Although starting off as a mystery series, it later evolved into secret agent and fantasy style plots. The use of drugs to render people unconscious appears in several episodes, in addition to illegal shipments of recreational drugs. There is also a scene in a photographer's darkroom in the episode *Interlude in Venice* (1966), with large chemical containers shown in the background as well as glassware. As Templar attempts to burn the photographs to protect the daughter of a friend, the photographer throws a small chemical container at him which smashes and releases a cloud of smoke.

In addition to chemicals, there are also two episodes featuring chemists. The first is *The Good Medicine* (1964), which follows a couple who build a large cosmetics empire. The episode tells the story of how the couple started with a *pharmacie* in the French town of Beauvais. The pharmacy is shown stocked with bottles and pills, as well as glassware and a mortar and pestle. It even shows the male chemist pouring a liquid from a beaker into a measuring cylinder at eye level to prevent parallax. Funnels are also used to fill containers to prevent spillage. The male chemist is described as "a top-notch chemist" but has "less sex appeal than the Sphinx". His wife encourages him to develop cosmetics, but later she ruthlessly kicks him out of the company and divorces him. To get some compensation from her, Templar fools her into investing in a new "insect repellent" pill.

Templar also meets a chemist in Islands of Chance (1967) while visiting the British West Indies. However, the meeting is brief since the chemist collapses and dies at the beginning of the episode. Templar discovers a small poison dart at the crime scene and meets a photojournalist who is interested in a famous local doctor and one time employer of the chemist. We are introduced to the doctor in a lab containing some condensers with tubes, a microscope, a burette, and a conical flask on a tripod over a Bunsen burner. There are also other items of glassware scattered around the room on benches.

The doctor explains that he is attempting to rid the world of all disease, through inoculation with a new "serum". He also explains that he works on the island because he is considered a murderer in the US for experimenting on a sick child with his serum. Later, Templar discovers that the chemist was killed because he discovered the stolen gold bars from a crashed airplane. When the henchman is asked where the gold came from, he jokes "Goldfinger, it came from Fort Knox". The composition of the poison is never explained, nor the serum.

6.6 RECREATIONAL DRUGS

Just like the space-race influenced sci-fi, real-world medical and recreational drug use also influenced movie themes. In *Seconds* (1966), a man decides to fake his own death and undergoes surgery to change his face. Using "pentothal and caffeine sodium benzoate", they determine what his deepest desires are for his new life. The chemical name for pentothal is thiopental, a barbiturate that was first developed as an anaesthetic but later found use as a 'truth drug' by psychiatrists. It works by placing the person in a semi-unconscious or hypotonic state, which can also make people susceptible to suggestions. Thiopental appears in numerous movies both as a 'truth drug' and as a 'brainwashing' drug. A combination of caffeine and sodium benzoate is also used to treat respiratory depression, a common side effect of thiopental.

Previously we discussed references to opium in *The Two Faces of Dr. Jekyll* (1960) and other recreational drugs in *Caprice* (1967).

In the same year we also have *Valley of the Dolls* (1967), opening with an image of three coloured pills spilling open, showing a white crystalline powder inside. Each of the pills (referred to as "dolls") represent one of three female characters navigating careers in the entertainment business, using "dolls" to keep up. The characters are played by Barbara Parkins, Patty Duke, and Sharon Tate. Many years later, Sharon Tate was played by Margot Robbie in *Once Upon a Time... in Hollywood* (2019), in which she watches herself in *Valley of the Dolls* (1967) at a Hollywood cinema.

The pills are referred to as "sleeping pills" and "pep pills" (most likely amphetamines), but their chemical makeup is never mentioned. When shown, only the characters' names are present on the labels, implying that they are prescribed. Eventually one of the characters, played by Patty Duke, has a psychotic break, stating "they say I'm drunk even when I'm not. Sure, I take 'dolls' because I got to get some sleep, I got to get up at 5 o'clock in the morning and sparkle". She is warned not to "take liquor with those pills", but she claims it helps them work faster. Alcohol does cause a slowdown in motor function, however mixing most pharmaceuticals with alcohol is always ill advised. She is also told by her agent that she looks 10 years older than she is due to the pills. Eventually, she "doesn't feel anything anymore" and ends up "taking an overdose". While recovering in the hospital she laments that "they say getting off them is worse than booze or dope. I'm scared – I've forgotten how to sleep without 'dolls', I can't get through the day without a 'doll'".

Next we have *The Trip* (1967), as we explore LSD (lysergic acid diethylamide), colloquially known as 'acid'. After a brief period of use in psychotherapy in the 1940s, LSD and other psychedelics were adopted by the counterculture movement due to their perceived ability to expand consciousness. A year after *The Trip* (1967) was released, LSD became a Schedule 1 controlled substance (illegal for medical and recreational use) in the US in 1968, followed by the UN in 1971.

The most relevant scene here is at the beginning, where three white pills are shown on screen and our protagonist is told that they contain "250 micrograms a piece". A microgram is 1 millionth of a gram, abbreviated as µg, which is correct for the dosage of LSD. However, 250 µg is considered a very large dosage. Next, he is presented with a pill of Thorazine just "in case you happen to go on a bad trip, this brings you back right away". In the real world, Thorazine is the brand name for chlorpromazine, an antipsychotic used to treat mental health disorders like schizophrenia and bipolar disorder. After taking one of the pills, he tells the dealer that the police raided the house of one of their friends the previous night after "finding her stash". The rest of the movie mainly concerns his trip.

We then finish with the iconic road-trip movie *Easy Rider* (1969), which opens with two bikers purchasing an unnamed white powder outside a Mexican bar. They test it by snorting, leading us to believe that it is cocaine. They bring it to Los Angeles where they sell it on to a chauffeur driver which provides the money they need for their road trip to Mardi Gras in New Orleans. Along the way they meet different people and take various drugs, but none are mentioned by name. One exception is a scene where the character played by Jack Nicholson is hesitant about smoking his first marijuana cigarette, saying "I might get hooked and it leads to harder stuff". After smoking the cigarette, he tells the group about UFOs, claiming that "they've been coming here ever since 1946, when the scientists first started bouncing radar beams off of the moon", which is called a "crackpot idea" by the others.

6.7 SALT JOKES ARE SODIUM FUNNY

The popularity of the James Bond movies throughout the 1960s inspired the first parody of the concept in *Our Man Flint* (1966). Our hero is tasked with stopping "mad" and "highly intelligent" scientists, who are blackmailing the world with a weather-control machine to force all nations to destroy their nuclear arsenal. When preparing for the mission he is presented with a suitcase of colourful packets labelled $C_7H_6O_3$, $H_3H_4H_6O_8$, and C_3H_4O. The first is the correct formula for salicylic acid, used in face cleansers, the second one doesn't make any sense, and the last one could be any number of real compounds, *e.g.* methoxyacetylene. Also, the chemical analysis of a poison dart comes back as "curare" and "in addition to the poison, there were traces of garlic, saffron, and fennel on the feather", which leads him to Marseilles. Curare is the common name for various real-world alkaloid arrow poisons used for hunting and it was also used as the first paralytic in anaesthesia. In the movie, the scientists always wear lab coats and represent an almost anti-hero depiction of 'mad scientists', who want to "end all kinds of poverty, disease, doubt and suspicion".

Next, we have the dark comedy *Dr. Strangelove, or: How I Learned to Stop Worrying and Love the Bomb* (1964), about an insane American general who orders a nuclear attack on the Soviet Union. At one point he claims that communists only drink vodka because the fluorination of water is "the most monstrously deceived and dangerous communist plot we've ever had to face", so he only drinks distilled water or "pure grain alcohol". In the real world, fluoridation became an official policy of the US Public Health Service in 1951 to prevent tooth decay. However, almost immediately there were numerous court cases arguing that it was "forced mass medication". Later in the movie, a Russian dooms-day machine is described as containing "cobalt thorium G" which will render the Earth's surface uninhabitable for 93 years. In the real world a mixture of radioactive cobalt and thorium does not result in a new single half-life and none of the known isotopes of thorium or cobalt have a half-life anywhere near 90 years.

The modern incarnation of the 'mad scientist' continues in the family comedy *Mad Monster Party?* (1967). This stop-motion animation opens in a castle lab where a scientist in a lab coat called Baron von Frankenstein is holding a colour-changing test-tube, which he incorrectly calls "a vial". He raises it up above the roof to get hit by lightning, "infusing it with energy" and creates a liquid which explodes like a nuclear bomb. Later, we also have a scene in a pharmacy with a clumsy assistant with round glasses stealing pills and drinking from a test-tube. Dr Jekyll also makes an appearance and changes into Mr Hyde, but the clumsy assistant thinks he is feeling seasick. Dracula also calls himself "the original Batman".

Finally, we finish this discussion with the first screen adaptation of the Caped Crusader in *Batman* (1966–1968), staring Adam West in the titular role. Throughout the TV series and TV movie, Batman performs quick chemical tests to analyse various substances. Usually, these tests provide him with a clue to move the plot along. In the very first episode, *Hi Diddle Riddle* (1966), we are shown an elaborate setup of glassware in the Batcave labelled "Chemistry Research Materials". In addition to a bench of glassware, there are also various flasks filled with a red solution and joined together with tubing. We will return to Batman again later, in addition to many other superheroes.

REFERENCE

1. R. E. Huschke, A Brief History of Weather Modification Since 1946, *Bull. Am. Meteorol. Soc.*, 1963, **44**(7), 425–429.

Public Enemy

7.1 THE WAR ON DRUGS

In 1971, the US President Richard Nixon famously declared a 'War on Drugs', with wide-ranging implications for many decades, especially for some communities. But this wasn't the first time new laws were enforced to control recreational drugs, with anti-opium laws enacted in the 1870s and anti-cannabis laws appearing in the 1910s and 1920s.[1] However, new drugs and their widespread availability after WW2 resulted in what psychiatrist Thomas Bewley called a "small epidemic of heroin and cocaine addiction".[2] However, much of the drug epidemic of this era was also generated by the pharmaceutical and medical profession, pushed on by widespread cultural shifts. Amphetamines were first discovered during the search for a decongestant in the late 1920s, but were later marketed and prescribed for depression. Their use as nonmedical (recreational) stimulants eventually spiralled out of control, reaching epidemic levels in the US by the end of the 1960s.[3]

Following the use of amphetamine pills in *The Valley of the Dolls* (1967), we have a similar depiction in *All That Jazz* (1979) where the protagonist, a stage director and dance choreographer, uses "pep pills" to keep up with the demands of show business. The movie opens with an image of a pill bottle with a

Onscreen Chemistry: The Portrayal of Chemical Science in Film and TV
By John O'Donoghue
© John O'Donoghue 2025
Published by the Royal Society of Chemistry, www.rsc.org

label that reads "Dexedrine 5 mg", which our protagonist takes as part of his morning routine before saying "it's showtime folks". Dexedrine is a real brand name for dextroamphetamine, a potent central nervous system (CNS) stimulant sometimes used to treat ADHD and narcolepsy. However, like other amphetamines, it is also used recreationally as a 'club drug' for its energetic and euphoric high. There are numerous side effects such as cardiovascular issues among others. In *All That Jazz* (1979), his morning routine is repeated a few more times onscreen before he eventually suffers a debilitating heart attack which later proves to be fatal.

We also have cocaine references in *The Seven-Per-Cent Solution* (1976) in relation to Sherlock Holmes's addiction. It opens with a suspicion that Holmes is abusing cocaine, with Dr Watson stating "Moriarty was a name I only known him to mutter when in the thrall of one of his cocaine injections". The camera focusses on one of Holmes's eyes showing a dilated pupil, which is commonly called 'cocaine eyes'. A Victorian syringe is also shown on a side table as we enter the room. Suspecting that he has suffered a psychotic break, Watson tricks Holmes into travelling to Vienna to seek treatment from Sigmund Freud. Later, Freud explains that he freed himself from "the power of cocaine" but, also, that one of his colleagues died from cocainism. Under hypnosis, Holmes admits that he started using cocaine at university, perhaps implying that universities are the gateway to drug use.

7.2 POLICE CHEMISTRY

Following on from the famous private detective, the 1960s and 1970s featured a plethora of movies and TV shows centred around detectives, *e.g. Dirty Harry* (1971), *The Long Goodbye* (1973), *Baretta* (1975–1978), *The Rockford Files* (1974–1980), and many more. The theme of fake and/or suspicious deaths was also frequently explored, as seen in *The Running Man* (1963), alongside recreational drugs.

We start this section with an unscrupulous doctor who is accused of murdering his wife as part of an insurance scam in *One On Top Of The Other* (1969). Early in the story, the protagonist opens the medicine cabinet and tells the new nurse that

"Listoprel" is used for his wife's asthma attacks. However, it is unclear what he is referring to. In the real world lisinopril is used for treating high-blood pressure, but one of its side effects is that it can make asthma worse. On the other hand, Isuprel (isoprenaline) is sometimes used for asthma attacks, so perhaps this is a mispronunciation as we have seen before in *Monkey Business* (1952). He goes on to explain that "before bed she takes 20 drops of Neurosedyl, it's a tranquiliser", but it is not clear what drug he is referring to here either. The closest name is Neurocil, which is a low-potency antipsychotic. He says that during an asthma attack "Neurosedyl can cause suffocation". Eventually his wife is found dead, leading to a police investigation.

There is a brief scene in a police forensics lab with several large chemical containers shown on a bench. A scientist in a lab coat shows the police some slides of signatures from a document, prompting them to exhume the body for an autopsy. This leads to a montage of laboratory glassware with round-bottomed flasks, condensers, conical flasks, test-tubes, bubbling solutions, and tubing with colourful liquids. Eventually they find an envelope of money with the protagonist's fingerprints on it. He is sentenced to death by gas chamber for killing his wife, but it later turns out she faked her death to frame him and switched places with her impersonator.

We then head to the Mediterranean for the Italian murder mystery *The Cat o' Nine Tails* (1971). The "nine tails" refer to the number of leads that the detectives need to follow to solve the crime. The crime revolves around a medical institute which is researching the genetic XYY syndrome (also known as Jacobs syndrome in the real world). The researchers are trying to predict the tendency for criminal behaviour from the presence of the "XYY chromosome", which prompts a killer to hide the evidence. In the institute lab, a scientist is shown using a separation funnel in a glove box with a measuring cylinder and some test-tubes. Volumetric flasks are also shown in the background containing brightly coloured liquids, as well as an inactive Bunsen burner, some retort stands and a microscope.

Later, as detectives search the institute, we get a brief look at all the labs at the institute. There are conical flasks with coloured liquids on benches as well as some empty test-tubes, chemical containers and other glassware in a realistic setting, *i.e.* no random tubes. In the real world there is no evidence for those with XYY syndrome to have increased criminal tendencies.[4] Nonetheless, *The Cat o' Nine Tails* (1971) is noteworthy for the controversial concept of predicting criminal behaviour (precrime), first coined in the 1956 Philip K. Dick novella *The Minority Report*, later adapted for screen as *Minority Report* (2002). The TV series *Law and Order* (1990–2010) also explores the concept of genetics linked to crime on several occasions.

In the same year we have *The French Connection* (1971), based on a true story about US detectives intercepting a large shipment of heroin from France. Early in the movie there is a display in the background of the police station with different types of crack pipes, pill bottles and bags. The key chemistry scene takes place in a hotel room when the US dealers want to test the purity of the heroin. A character who is never referred to as a chemist, only assumed, pours some mineral oil through a funnel into a Thiele tube, which is then fitted with a thermometer and a small sample of the drug. Using a spirit burner, the temperature rises until it eventually tops off at 240 °C, at which point the character says "89% pure". Since the boiling point of pure heroin is 272–274 °C in the real world, he is basing the purity here on the fact that 240 °C is 89% of 272 °C.

A Thiele tube is a piece of laboratory glassware designed to contain oil and is normally used for determining the melting point or boiling point of a substance. Since the melting point of heroin is 173 °C (343 °F) and impurities would only deviate this by a few degrees, our hotel chemist is using the boiling point to determine the purity instead. It appears that he is using the Siwoloboff method, which involves attaching a sample in a capillary tube to a thermometer using a rubber band and placing it in an ignition tube over an oil bath such as in a Thiele tube. But this is normally used for liquids, not solids as shown here. Notwithstanding this, the attention to detail is very impressive.

However, this does result in a negative association of chemistry skills and knowledge of illegal recreational drugs, in addition to the fact that the chemist is paid for his services with a bag of heroin. Later, when the heroin is removed from the car, it is confirmed by the chemist with a few drops of an unnamed liquid, turning it from white to violet.

Following this, we have a few interesting scenes in the *Kojak* (1973–1978) episode *Mojo* (1974), where Kojak (Telly Savalas) goes undercover as a chemist to investigate the theft of powdered morphine sulfate from the Dexter Pharmaceutical Co. First, we have a lab scene with pipettes and test-tubes, where a chemist explains the morphine test to Kojak as "some sulfuric acid should do it. If it's morphine it will turn a deep violet". This is true, but only when it is heated. This can also be used for heroin when mixed with formaldehyde (Marquis reagent), which is most likely the test shown at the end of *The French Connection* (1971).

When he meets the thieves to run the test, Kojak takes some test-tubes, a clock glass and a sample tray out of his bag to analyse the morphine. He then uses a small pipette to drop a liquid onto the white powder, which turns dark pink. He explains that this test "only indicates the presence of morphine, not its purity. We must make sure it is not diluted". When asked how long the next test will take, he exclaims "you can't rush chemicals, sir!". He then heats a sample over a flame and attempts to use the weight of the residue with a small printed table to determine its purity. Onscreen we are shown a list of elements along with their correct atomic weights and common oxidation numbers. But morphine is also included on this list alongside the symbol for mercury (Hg) and the approximate weight doesn't make any sense. But since Kojak is only pretending to be a chemist to fool the thieves, I think we can forgive him for this.

Next, we have the first proper depiction of forensics onscreen since *Kid Glove Killer* (1942) in *Hawaii Five-O* (1968–1980). The lab at the police headquarters is run by the multi-talented forensic scientist Che Fong (Harry Endo), who provides a core function to the plot in dozens of episodes. It's fair to say that every application of chemical forensics is covered at some point during this long-running TV series. It's very likely that the longevity and popularity of the show later inspired the plethora of 'crime scene investigation' themed TV series in the early 2000s. It was also later rebooted as *Hawaii Five-0* (2010–2020), but this updated version downplays the forensics in favour of medical examinations instead.

The lab in the original series is frequently shown with test-tubes, chemical containers, and a microscope, but activity is usually brief in the early seasons. Later, there are several episodes based around common chemistry themes like medicines in *Air Cargo – Dial for Murder* (1971), where the team investigate stolen pharmaceuticals. The drug in question is glucagon, a real-world hormone which increases blood glucose levels, *i.e.* the opposite of insulin. But there isn't really any analysis carried out, only a brief scene in the lab with chemical containers in the background as they trace an automated phone service.

Away from drugs, Che also looks at fragments from explosives like "the outer wrapping of a stick of dynamite" in *Engaged to Be Buried* (1973). After dipping the fragment in a beaker of red liquid and looking at it under a microscope, he uses a table of references to identify a trademark on a label. Similarly, he also provides information about gunshot residues on clothing and about fire debris to determine the cause of arson. He also analyses handwriting, ink, and fingerprints in conjunction with the local university, as seen in *The Ninety-Second War* (1972). There are also occasional references to pollution and water analysis, as well as hazardous chemicals. It should be noted also that Che isn't confined to the lab: he also visits crime scenes to collect evidence and his tool of choice is usually a lab tweezers.

But the most significant episode from a chemistry perspective is *Nine Dragons* (1976), which centres around toxins and a university chemistry department. It features numerous scenes in a chemistry lab, as well as chemists who are under strict orders about storing the toxins while "the experiments are being carried out". There is also a description of an "environmental vacuum chamber" which "has allowed us to study physical matter in a new and undistorted way". Eventually the toxins are described as cobra venom and saxitoxin, a real Schedule 1 chemical weapon derived from shellfish. However, despite using a pair of tongs to handle the metal containers, the chemist then uses his bare hands to open them on the bench and remove the glass vial. During the experiments we get a brief look at a test-tube being heated over a Bunsen burner.

Later, during the process of stealing the toxins, all the chemists are rendered unconscious and Freon-12 gas is discovered attached to the regular gas line. Freon is a refrigerant used in older forms of fridges and air-con systems, now banned along with other chlorofluorocarbons (CFCs) due to the damage they cause to the ozone layer. When burned, like in the episode, it produces phosgene, a real Schedule 3 chemical weapon used during WW1. Finally, there is a brief mention of cinnabar (mercury sulfide), which is used in a drink as part of an initiation ceremony for the Nine Dragons triad gang.

Overall, the level of chemistry onscreen is impressive throughout the series and sets a high bar for future TV forensics. Hawaii Five-O (1968–1980) was replaced by two other famous police shows, but forensics did not feature in any significant way. Drug smuggling appears in *Magnum PI* (1980–1988) from the beginning with *Don't Eat the Snow in Hawaii* (1980) and again in other later episodes. However, cocaine and occasionally heroin featured prominently in the US crime series *Miami Vice* (1984–1989), with some scenes taking place in illegal drug processing labs.

Finally, there is one more TV series that needs a special mention here. *MacGyver* (1985–1992) follows the adventures of the physics-trained secret agent Angus MacGyver (Richard Dean Anderson). It famously features his resourcefulness and knowledge of the physical sciences to solve complex problems using ordinary objects. Interestingly, MacGyver's exploits were vetted by consulting scientists for accuracy. Importantly though, where MacGyver used household chemicals to create poisons or explosives, specific details were deliberately omitted, altered, or left vague, in the interest of public safety.[5]

In the first season, most of the chemistry is associated with making explosives. He calls one of his explosive creations "kitchen chemistry" in the episode *Every Time She Smiles* (1986), using a pesticide, soap flakes and "tile cleaner". The detonator consists of lard covered with newspaper and cooking oil, placed in a gas oven. Also, in the episode *Eagles* (1986), he creates a "low-level spontaneous combustion" from an egg incubator, chair padding and vegetable oil. Finally, he uses nitrate from fertilizer, cellulose from plant bark, and "a few drops of acid" to make a nitromannite bomb to blow up a dam in the episode *Trumbo's World* (1986).

But the chemistry isn't all explosions: he also uses vinegar and baking soda to create a CO_2 fire extinguisher in *Good Knight MacGyver* (1991). Similarly, he also makes a spectroscope from nail-polish remover, a lamp, and polarized lenses to investigate the use of illegal pesticides in *Bitter Harvest* (1990). From the spectroscopic lines produced, he is somehow able to tell which pesticide it is, which is highly unlikely. Finally, he also makes a battery from a galvanised nail and a copper coin in a cactus to power a radio in *Ugly Duckling* (1986). Although this is technically possible, this only represents one battery cell – he would need many more to produce enough current to power a radio. The popular series was later rebooted as *MacGyver* (2016–2021).

7.3 DYSTOPIAN ACCIDENTS

After the optimistic 1950s and brightly coloured 1960s, stark depictions of dystopian futures began to emerge in the 1970s. Before we begin this discussion, it is worth noting several prominent industrial accidents in the real world at the time, such as the US Thiokol-Woodbine explosion (1971), the UK Flixborough Nypro chemical plant explosion (1974), and the Italian Seveso ICMESA dioxin chemical contamination (1976). Industrial accidents and growing environmental concerns are credited as the inspiration for *The Omega Man* (1970), *Silent Running* (1972), *Logan's Run* (1976) and *Mad Max* (1979), among many others.

First in our discussion here is *THX-1138* (1971) by George Lucas, about a future where sexual intercourse and reproduction are prohibited. Mind-altering drugs are used to enforce compliance, supress emotions and allow dangerous tasks to be completed, despite frequent accidents, *e.g.* handling radioactive material. In one scene, the sedative drug is mentioned as "Etrazene". The name and function of this substance has similarities to the real-world Schedule 1 controlled substance etodesnitazene, which is sometimes abbreviated to etazene. Discovered in the late 1950s, this opioid analgesic is seventy times more potent than morphine. Various pills are also shown throughout and there are multiple references to "chemical imbalance" when drugs are switched.

The dystopian chemistry continues in a densely overpopulated and starving New York City (NYC) of the future, where a detective investigates the murder of an executive at a rations food manufacturer in *Soylent Green* (1973). Set in the year 2022, the population of NY is said to be 40 million. The estimated population of NYC in the real 2022 for the entire metropolitan area is estimated to be about 23 million (it was about 16 million in the 1970s). Our protagonist's roommate is a former professor who tells him what food was like before "our scientific magicians poisoned the water and decimated the soil" resulting in "a heatwave all year long, the greenhouse effect". All outdoor scenes in the movie are presented with a green tint throughout as a reference to the new food item.

Interestingly, electricity in their apartment comes from a bike attached to a generator which charges a bank of batteries in the background, something that is now familiar in the form of domestic solar power systems. But the worn-looking batteries are not very good at holding energy, since the lights fade shortly after they finish cycling. The titular food product Soylent Green has just been launched and is marketed as a derivative of plankton. Later, our protagonist discovers that an executive from the company was an expert in freeze drying and was murdered because he discovered a conspiracy involving the Soylent company.

Next, we have *Westworld* (1973), later remade and modernised as a TV series (2016–2022), about autonomous androids who go out of control in a futuristic theme park. There are some brief scenes with glassware filled with solutions in the original movie when the androids are being repaired in the labs. In one scene we are shown acids on a tray with labels containing the correct chemical formulae "Nitric Acid HNO_3", "Hydrochloric Acid Con. HCl", "Sulphuric Acid Con. H_2SO_4" and "Acetic Acid $HC_2H_3O_2$". The last one is more usually written as CH_3COOH to better reflect the molecular structure, but it is still correct here. The protagonist picks up the concentrated HCl and throws it at the Gunslinger android, which partially melts his face in a cloud of smoke.

HCl can cause blindness and severe chemical burns on human skin in the real world, but we don't know what the android synthetic skin is made from. HCl doesn't normally react in the same way with silicon or rubber-based materials. However, the term 'artificial skin' was first used in the early 1970s to describe treatments for fire burns, consisting of a collagen scaffold, which would be more sensitive to HCl.[6] Finally, of note here also is the fact that the scientists/technicians are always wearing white coats, even in meetings, almost as a way of immediately identifying and contrasting them from the androids and guests of the resort.

Next, we have one of the first movies about a chemistry-based industrial accident, *One Man* (1977). This Canadian drama follows a journalist investigating a chemical leak which has poisoned several children. We later discover that the cause is "BAP – biacetyl plumbane", which the doctor describes as a "cumulative poison, like lead or mercury". BAP is referred to as a gasoline additive which has been spread everywhere due to vehicle exhaust fumes. People are afraid to speak out about the source of BAP due to intimidation and threats of violence. After obtaining a report, our protagonist is attacked by thugs hired by the manufacturer of BAP under the guise of union workers protecting their jobs. Through obsession and perseverance, the journalist eventually confirms that it was the chemical leak which caused the deaths of the children, despite resistance from the company.

In the real world, BAP refers to lead(II) acetate (old name plumbous acetate), which was historically used as a sweetener and preservative in wines before its toxicity was discovered.[†] Here, it is used as an analogy for the real gasoline additive called tetraethyl lead (TEL) which increased the octane rating, performance, and fuel economy of petrol engines. Due to the toxic effects of lead, especially on children, it was phased out from the 1970s onwards. The word 'plumbane' refers to the Latin name for lead, which is also the source of its elemental symbol.[‡]

[†] The earliest confirmed poisoning by lead acetate was that of Pope Clement II who died in AD 1047. A toxicological examination of his remains conducted in 1959 confirmed centuries-old rumours that he had been poisoned with lead sugar (lead(II) acetate or plumbous acetate).[9] The toxicity of the real-world gasoline additive tetraethyl lead (TEL) was played down by oil and car companies for many decades since the additive had become essential to avoid knocking in petrol engines.[10] The toxicological examination of Pope Clement II was published 18 years before *One Man* (1977), so it is likely that BAP is referring to lead sugar as a way of implying that the toxicity of lead compounds had been long known.

[‡] The Latin word for lead is *plumbum*, providing the elemental symbol Pb. The Romans used lead to make water pipes due to its malleability and durability, leading to the words 'plumbing', 'plumber' and 'plumb bob'. The latter is a weight traditionally made from lead and used to keep a string taut as a reference (plumb line), previously mentioned in relation to *Journey to the Center of the Earth* (1959). 'Plumbane' also refers to lead hydride (PbH_4), a thermodynamically unstable compound that is difficult to isolate.

In contrast to movies which hint at real-world toxins, there are also examples of movies predicting the future, like the flat-screens and tablet computers in *2001: A Space Odyssey* (1968). However, the most incredible prediction is from *The China Syndrome* (1979), released in US cinemas only 12 days before the real-world US Three-Mile Island accident (1979). The name is derived from the exaggerated result of a nuclear meltdown, where the reactor components melt through their containment "all the way to China". The movie portrays a coverup of safety concerns at a fictional nuclear power plant by an 'evil industry' (utility company) who are more concerned about losing money than the safety of the plant.

A pellet of uranium is described as having more energy than six carloads of coal. But it is also mentioned that uranium is used in atomic bombs. One of the few slips in terms of accuracy involves an explanation about how the reactor works, which is said to boil water to make steam to spin a turbine. However, the illustration shows a steam generator between the reactor and the turbine, meaning it is a pressurised water reactor, only boiling water reactors produce steam directly. Despite the name of the movie, a meltdown is prevented because the safety systems (SCRAM) kick in due to a turbine trip caused by a malfunctioning pressure relief valve and water level indicator. The events in the movie are strikingly similar to the real-world Three Mile Island accident which suffered a partial meltdown.

7.4   THE CREATIVE LAB

Despite the drift towards recreational drugs and industrial accidents, we still have some examples of chemistry inventions during this era. The first is the British movie *Willy Wonka and the Chocolate Factory* (1971) which returns us to a school chemistry lab for the first time since *The Belles of St. Trinian's* (1954). The chemistry teacher, amusingly named Mr Turpentine (David Battley), asks the protagonist, Charlie (Peter Ostrum), to help him with a demonstration. Mr Turpentine explains that he has "nitric acid, glycerine and a special mixture of my own. Together, it's horrible, dangerous stuff – blows you up". Behind them on the blackboard is "H–O–H" and on the shelf are some test-tubes and other glassware, as well as several chemical containers.

The authoritarian teacher belittles Charlie by explaining that only the teacher knows the correct mixture. He claims that the three ingredients mixed in the "right way" create the "finest wart remover in the world". They then pour all three bottles into a bucket at the same time, causing a small explosion with lots of smoke, returning us to the 'chemistry causes explosions' trope. Mr Turpentine then dismisses the class when he receives news about Wonka opening the factory. We return to the teacher later who uses the Wonka bars and the golden ticket to explain percentages, but once again picks on Charlie and mocks him for only opening two Wonka bars compared to the large numbers consumed by his classmates. Overall, these scenes represent a very negative image of chemistry teachers.

Interestingly, *Willy Wonka and the Chocolate Factory* (1971) also depicts the concept of underhanded industrial rivalry in the form of the antagonist, Arthur Slugworth, who uses bribery to convince people to steal Wonka's invention "the everlasting gobstopper". During the exploration of the chocolate factory, we are shown "the inventing room" which, in addition some chaotic machines, contains lab glassware and a Bunsen burner. Wonka (Gene Wilder) uses the glassware to mix some liquids and explains "invention, my dear friends, is 93% perspiration, 6% electricity, 4% evaporation, and 2% butterscotch ripple", which one of the parents points out is 105%.

Next, we have the *Dexter Riley* series of TV movies, a spin off from *The Absent-Minded Professor* (1961) and *Son of Flubber* (1963), featuring the same college. The first movie, *The Computer Wore Tennis Shoes* (1969), sees Dexter (Kurt Russell) gain superhuman maths talents. Next, we return to the concept of invisibility and college chemistry labs in *Now You See Him, Now You Don't* (1972). It opens with an interesting debate about chemistry teaching, when the professor requests new lab equipment from the Dean because "our students have to experiment, be creative, now that's what science is all about today". This is contrasted with the fiscal Dean who tells the professor to "forget about your course in creative lab. Just stick to the old conventional systems where you burn the sulfur, make the clouds, and smell up the place like you do". The professor replies "that was only good enough for the 1940s".

The chemistry lab is presented with plenty of glassware set up for reflux filled with colourful solutions. In a throwback to *The*

Chemist (1936), one student appears to be doing chemistry but is actually adding milk to cereal, which he then proceeds to eat in the lab. Dexter has an elaborate setup of colourful glassware, lots of tubing and sparks of electricity. He tells the Dean that "a lot of things have happened since Einstein split the atom", which is a common mistake. Einstein laid the theoretical foundation for understanding nuclear reactions and warned US president Franklin D. Roosevelt about it in a letter. However, nuclear fission was first achieved by the British and Irish physicists John Cockcroft and Ernest Walton in 1932.

Dexter also talks about a Russian scientist from 200 years ago who was working on invisibility, which is most likely referencing the real-world Soviet engineer Pyotr Ufimtsev. Ufimtsev is considered to be the 'father of stealth' for his theories on the scattering of electromagnetic waves, and his work was translated into English in 1971. Later, in a homage to *Frankenstein* (1931), a lightning storm strikes the roof of the lab, causing Dexter's experiment to boil and change colour. It creates an invisibility solution which, unlike previous examples, can be washed off, setting up the antics for the movie. One use for the invisibility is helping the Dean win at golf, which represents another rare chemistry and sport crossover.

We return to Dexter and the 'creative lab' in *The Strongest Man in the World* (1975) with a formula for super-strength. The lab is presented with a similar setup as the first movie, but with more colour and tubing. However, the correct chemical formula for sodium carbonate can be seen on the chalkboard as Na_2CO_4. From a safety point of view, most of the glassware, filled with colourful solutions, is clamped dangerously high off the bench. There are also several examples of tasting and eating in the lab, like before. An accidental mixing of solutions is the cause of the superhuman formula, which also powers up a car. We also have another chemistry and sport crossover, this time in relation to weightlifting. Later, a fight breaks out in the lab, which results in a lot of broken glassware. Several examples of fight scenes and destroying glassware will be discussed in greater detail later.

Interestingly the overall theme of the Dexter Riley series is financial troubles in the college, which is so bad that they even accept funding from a known criminal. The science department is usually blamed for the cost overruns, and they are told that they need to invent something that makes the college money. Note, there is no known connection between the *Dexter Riley* movies and the animated TV series *Dexter's Laboratory* (1995–2003), although perhaps the animated series was inspired by these movies.

7.5   MONSTERS AND ALIENS

First here we have *The Andromeda Strain* (1971), about a team of scientists investigating a deadly extraterrestrial organism (andromeda) which they extract from a black and green meteorite. The scientists are introduced wearing sealed environmental suits complete with oxygen tanks as they investigate the deaths of a large number of people. They are then transferred to a large underground facility which contains a "thermonuclear self-destruct system". Other than a very elaborate decontamination procedure, the movie centres around the use of advanced technology, which includes state-of-the art chemistry instrumentation for the time period. The scientists also demonstrate extreme determination and perseverance as they work around the clock to figure out how to stop the andromeda organism.

The first instrumentation shown is in the microchemistry lab where the scientists are using a mass spectrometer to analyse the black object. It returns an elemental percentage with 21.07% hydrogen, 54.90% carbon, 16.00% oxygen, 2.00% chlorine, 1.01% sulfur and 0.20% silicon. From this they conclude that it isn't a rock, it is a "material similar to plastic". The green object comes back with 27% hydrogen, 45% carbon, 5% nitrogen and 23% oxygen. It's not clear why two decimal points are shown for the first sample, while only whole number percentages are used for the second. Next, we are shown inside the electron microscopy lab where they take an "800 ångströms thick" slice of the green object for analysis. From this they discover that andromeda is a crystalline organism which can "live on anything", without amino acids. It functions like "a nuclear reactor", which makes them realise that they need to turn off the self-destruct system.

Finally, we are also shown an X-ray crystallographic image of andromeda which proves that it has complete symmetry. They eventually discover that andromeda can only grow in a very narrow range of neutral pH, so if a person's blood is too acidic or too alkaline, it cannot survive. They advise one of the infected scientists to breathe rapidly (hyperventilation) to make himself go into "respiratory alkalosis", causing the carbon dioxide levels in his blood to drop. They apply the same theory to the outside world: "we're seeding the clouds above andromeda with silver iodide. The raindrops will carry the organism into the ocean, and the alkaline reaction from seawater should kill it." Overall, *The Andromeda Strain* (1971) is a fantastic example of how to use real-world chemistry and instrumentation to work through a problem step by step. It also proves that real and accurate chemistry can be used in conjunction with fictional stories.

Next, we return to the good doctor again, this time appearing as *Dr. Jekyll and Sister Hyde* (1971). Breaking from the usual story, this time the male Dr Jekyll transforms into a female Mrs Hyde (which he names after London's Hyde Park). The story also incorporates aspects of the real-world Jack the Ripper as well as the Burke and Hare grave robbers. Here Jekyll wants to cure all known illnesses, but he is mocked for how long his experiments take. Instead, he decides to look for the elixir of life using female hormones from cadavers provided by Burke and Hare. His theory is that these hormones will extend his life since women tend to live longer than men. But it is not explained why the women need to be dead to obtain the hormones. Obsession with the research appears again, distracting Jekyll from a woman who has affections for him. However, in a complete departure to the original concept, both Jekyll and Hyde could be considered evil here, since both murder women for their hormones.

In keeping with the Victorian setting, the lab is located within the house this time, instead of the basement. However, there is significantly less glassware compared to previous versions. Chemical containers are shown on the shelves in the background and some glassware is clamped in a sporadic fashion on his lab bench, connected with black tubing. Initially some of the experimental setup looks much more restrained and sensible. However, later we have an elaborate setup of glassware filled with colourful boiling solutions, all connected with extensive tubing. We do have rare examples of experimental skills though, when Jekyll removes a condenser from the top of a reaction vessel to add some blue liquid and when he uses a pipette to extract a small sample for testing. The mirror motif also appears again briefly, as a subtle reference to molecular chirality as discussed previously.

In the same year we also have a more traditional adaptation of Jekyll's story in *I, Monster* (1971), with the names changed to Dr Marlowe and Mr Blake. It opens in a small home lab with an elaborate glassware setup containing bubbling solutions and extensive tubing. There is an interesting scene where Marlowe pours a colourless liquid from a test-tube into a beaker, changing the purple solution into a white precipitate. Although not explained, this is most likely a classic chemistry demonstration involving the reduction of potassium permanganate from the +7 oxidation state to the +2 oxidation state.[7] Similar to the 1960s version, syringes are prominent here when Dr Marlowe experiments on himself and others with his intravenous drug which "releases inner inhibitions". The mirror also appears frequently, as well as animal testing (and animal cruelty!).

There is also a brief return to mutations featuring onscreen chemistry in this era with *Phase IV* (1974), about the rapid evolution of ants after a mysterious cosmic event. The scientists set up a lab near the ant colony to perform some experiments using advanced computer systems. There is only one scene featuring a chemistry lab, which mainly consists of tubes joining glassware together on a bench. One of the characters smashes the glassware containing the ants in retaliation for the ants killing her horse. This results in the lab being quarantined and fumigated, but eventually the ants take over the world.

Next, we have an example from the blaxploitation genre of movies, This was a genre designed to attract new black moviegoing audiences, with black characters as protagonists, rather than sidekicks, supportive characters, or victims of brutality. Following the success of *Blacula* (1972), we have *Dr. Black, Mr. Hyde* (1976), about an accomplished and wealthy African American medical doctor working on a cure for liver cirrhosis. There are several scenes in a realistic medical research lab with glassware such as beakers, conical flasks and test-tubes, in addition to electronic equipment. The use of syringes, animal testing and the mirror continue from other versions of the Dr Jekyll and Mr Hyde story, as well as the usual themes of perseverance and obsession. The protagonist (Dr Henry Pride, played by Bernie Casey) also conducts unethical experiments on himself and others, but this time it turns him into a white-skinned monster with superhuman strength and invincibility. Cocaine is also shown onscreen and used by some characters.

Finally, we return to the school lab, for the only the second time in the 1970s, in the Japanese supernatural horror *House* (1977), about a schoolgirl and her friends who travel to her aunt's haunted house. It opens with the girl taking photos of her friend behind some lab glassware which is producing smoke-like condensate. When asked why she was taking the photo, she replies that she looks like "a witch in a horror movie". As the scene zooms out, we see that they're in a school chemistry lab complete with test-tubes and round-bottomed flasks filled with colourful solutions. There are no explosions or tubing, but the association of lab glassware with horror is an interesting reference to the fact that onscreen chemistry was dominated by gothic horror stories for so many decades.

7.6   BACK BONDING

We now return to the famous British spy, with Sean Connery returning to spy duties in *Diamonds Are Forever* (1971). This time we open in a medical facility with some IV drips and tubing. Later, Bond describes diamonds as "the hardest substance found in nature", leading him on a journey to track down diamond smugglers. He later discovers a secret research facility, filled with people in white coats who warn him to be careful of radiation. He then commandeers a moon rover from a fake moon set to escape and we discover that Blofeld has built a space-based laser weapon using diamonds for refraction. Diamonds do have an exceptionally high refractive index (RI) compared to other minerals, but although RI is a significant variable, there are many other factors which dictate the power of a laser.

Next, Roger Moore takes over Bond duties in *Live and Let Die* (1973). Released during the blaxploitation era, it includes the first African-American 'Bond-Girl' (Gloria Hendry) and villain (Yaphet Kotto), but it also includes numerous clichés as well.

Notable for this discussion is the departure from megalomaniac supervillains (Blofeld), focussing instead on drug trafficking. We also have the first exaggerated gadgets in the form of a wrist-watch with a "hyper-intensified magnetic field, powerful enough to even deflect the path of a bullet" and a gun that shoots "compressed gas bullets". Both of which are examples of extreme exaggeration of some real-world concepts.

The villain (Dr Katanga/Mr Big) is growing vast poppy fields for heroin and plans to distribute it free of charge through his restaurants in the US, increasing the number of addicts. This will bankrupt other drug dealers, allowing him to cash in later. The lab used to extract the heroin is only briefly shown in the background with people in lab coats working on a conveyor belt. We can also see a balance, volumetric flasks and other glassware filled with colourful solutions. Bond eventually foils the plan and destroys the poppy fields.

In the middle of the real-world oil crisis, we then have *The Man with the Golden Gun* (1974) about a "Solex Agitator". The solex is described as a type of "solar charge controller" which is "95% efficient". The standard maximum power point tracking (MPPT) controllers now in widespread real-world use are over 95% efficient. They were first released in the 1980s, but were investigated for many years before this.[8] So the concept at the heart of the story is based on real-world research at the time. But, as is the norm for Bond movies of this era, the level of scientific accuracy fluctuates. Early in the movie, Bond brings the remains of a bullet to the lab for analysis, who tell him that it's made from "23 carat gold with traces of nickel". However, only a balance and some microscopes are shown on screen, so it is not clear how they came to this conclusion.

Bond later receives a tour of the "solar energy station" which strangely uses "thermo-electric generators to convert solar energy into electricity". The energy is stored in "super-conductivity coils, cooled by liquid helium", a real invention by Ferrier from 1970, and the temperature is correctly given as -453 °F. Large mirrors reflect the sunlight into a laser, creating "heat for the thermo-electric generators". Later, the security officer falls into the liquid helium, causing the plant's temperature to spiral out of control and explode. Interestingly, this is one of the most prominent examples of cryogenic liquids shown on screen during this era. Cryogenics was beginning to gain interest among the public at the time, particularly after James Bedford became the first person to be cryopreserved in liquid nitrogen in 1967.

We then return to the threat of nuclear war in *The Spy Who Loved Me* (1977), but there is no real onscreen chemistry to discuss. Then, two years after *Star Wars* (1977), three years after the completion of the first real-world US space shuttle (*Enterprise*), and over a decade after *2001: A Space Odyssey* (1968), we fly to *Moonraker* (1979). Here, the villain, Drax (Michael Lonsdale), plans to use a "nerve agent" to kill most humans and use the Moonraker space shuttles as transporters to save a chosen "super race" aboard a hidden space station.

After a brief scene in a glassblower's workshop, the lab is shown with scientists in medical coats using remotely operated robotic arms to handle vials of colourless liquid. There are also some chemical containers on shelves in the background. Later, analysis of the vial reveals an "element breakdown" in the form of a completely nonsensical, partially disconnected chemical structure with an incredible number of chemistry errors (Figure 7.1). Bond describes it as "the chemical formula of a plant", a very rare (fictional) orchid which causes sterility. But Drax has manipulated the plant into a nerve gas that only kills humans.

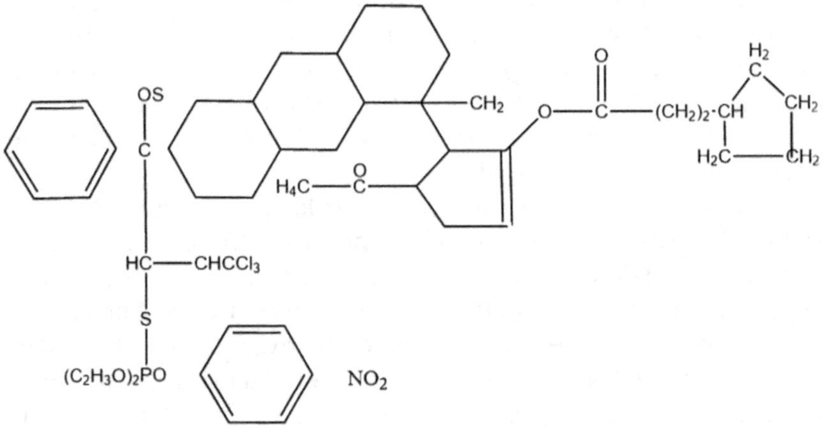

Figure 7.1   The "element breakdown" of the nerve gas as seen in *Moonraker* (1979). © Eon Productions, Les Productions Artistes Associés, United Artists.

As the Moonraker shuttles lift off, there are sensible references to liquid hydrogen and oxygen as real rocket fuel, but helium is also mentioned here for some reason. There are also several examples of lasers, such as a laser pistol that can melt objects and a laser cannon on the space station to destroy other spacecrafts. The movie cumulates in a space battle between US and enemy astronauts shooting laser-based guns in scenes that could easily be mistaken for *Star Wars* (1977). We finish with Bond and the CIA agent "attempting re-entry".

7.7 YOUNG CHEMISTRY

So far, we've only mentioned some examples of animations. However, with the increased availability of TV sets from the 1950s and 1960s onwards, the number of depictions of animated onscreen chemistry also increased. The famous titular Great Dane in *Scooby Doo* (1969–present) brought mystery solving to a younger audience during this era, along with a team of four teenagers. Occasionally, the series included similar themes to those discussed previously such as toxic waste, poisons, explosions, and a chemical company called Happy Sunshine Chemicals that goes radioactive and makes the town inhospitable.

One episode also includes a toxic chemical spill which creates a mutant who terrorises a town. There are also homages to Frankenstein and other gothic horrors with one such lab described as a "Mad Scientist Rumpus Room". In the same lab, Scooby and Shaggy start playing around with beakers and test-tubes as they try to make a formula to "turn werewolves into pussy cats". When the beaker starts bubbling, they call it a "reject" and throw it behind, accidentally causing a large explosion.

Next, we have *The Muppets* (1955–present), first created by Jim Henson in 1955. However, Dr Bunsen Honeydew didn't appear until 1976, followed by his long-suffering assistant Beaker in 1977. His first name comes from the real chemist Robert Bunsen, namesake of the Bunsen burner. His last name is a reference to the honeydew melon due to the shape of his head. He is depicted as a bald, yellow-skinned, bespectacled and lab-coated scientist with no eyes. He is always wearing a lab coat and the lack of eyes gives him the appearance of a type of absent-minded intellectual.

The experiments depicted onscreen usually go astray and cause harm to his assistant Beaker, who is of course named after the well-known piece of laboratory glassware. The lab scenes are usually a homage to the inventive nature of onscreen depictions of chemists from the 1950s and 1960s, with some examples including a hair growing tonic, a banana sharpener and an electric nose warmer, among many others. From a chemistry perspective, he also creates a device which references the ancient goal of alchemy, turning gold into cottage cheese. Beaker on the other hand wears a greenish lab coat and is usually depicted as shy. He doesn't speak any real words and only communicates in "meep" sounds. Over the years he has been blown up, electrocuted, eaten by monsters, and suffered several side effects due to Bunsen's experiments, all of which reference various movies from the golden era and the space age. Overall, I'm not sure what the writers are implying about those who run labs and their subordinates, but I certainly wouldn't want to be Bunsen's assistant!

Interestingly though, Bunsen and Beaker were not the first chemists to appear on *The Muppets* (1955–present). Worth a brief mention here is a musical sequence called *Time in a Bottle* from *episode 207* (1972). This one-off scene opens with an elderly chemist mixing colourful liquids and adding a white solid to a beaker. He is shown in a lab surrounded by other glassware like conical flasks, round-bottomed flasks and test-tubes, all of which are filled with colourful and bubbling liquids. He then drinks the concoction, which makes him younger with each verse of the Jim Croce song *Time in a Bottle*. However, at the end of the sequence, he reverts to his initial age with an explosion. As a result, his self-experimentation is unsuccessful and not even the wonders of chemistry can reverse the aging process.

REFERENCES

1. J. Hodge and N. Dholakia, Fifty Years Ago Today, President Nixon Declared the War on Drugs, Vera, 2021, https://www.vera.org/news/fifty-years-ago-today-president-nixon-declared-the-war-on-drugs.
2. A. Mold, Illicit Drugs and the Rise of Epidemiology during the 1960s, *J. Epidemiol. Community Health*, 2007, **61**(4), 278–281.

3. N. Rasmussen, America's First Amphetamine Epidemic 1929 –1971: A Quantitative and Qualitative Retrospective with Implications for the Present, *Am. J. Public Health*, 2008, **98**(6), 974–985.

4. NORD – National Organization for Rare Disorders, *XYY Syndrome*, Rare Disease Database, https://rarediseases.org/rare-diseases/xyy-syndrome/.

5. W. Britton, *Spy Television*, Bloomsbury Publishing, USA, 2004.

6. I. V. Yannas and J. F. Burke, Design of an Artificial Skin. I. Basic Design Principles, *J. Biomed. Mater. Res.*, 1980, **14**(3), 339.

7. C. O'Driscoll, N. Reed and T. Lister, *Classic Chemistry Demonstsrations: One Hundreds Tried and Tested Experiments*, The Royal Society of Chemistry, 1995.

8. M. Abdel-Salam, M.-T. EL-Mohandes and M. Goda, History of Maximum Power Point Tracking, in *Modern Maximum Power Point Tracking Techniques for Photovoltaic Energy Systems*, ed. A. M. Eltamaly and A. Y. Abdelaziz, Springer International Publishing, Cham, 2020, pp. 1–29.

9. W. Specht, K. Fischer, W. Katte, S. Berg and H. Hrabowski, Vergiftung Nachweis an Den Resten Einer 900 Jahre Alten Leiche" [Evidence of Poisoning in the Remains of a 900-Year-Old Corpse] (in German), *Arch. Kriminol.*, 1959, **124**, 61–84.

10. B. Kovarik, A Century of Tragedy: How the Car and Gas Industry Knew about the Health Risks of Leaded Fuel but Sold It for 100 Years Anyway, The Conversation, 2021, https://theconversation.com/a-century-of-tragedy-how-the-car-and-gas-industry-knew-about-the-health-risks-of-leaded-fuel-but-sold-it-for-100-years-anyway-173395.

Accident Prone

8.1 ACCIDENTAL CHEMISTRY

Previously we saw how the devastating industrial accidents of the 1970s were beginning to have an impact on depictions of chemistry onscreen. However, the real-world accidents continued into the 1980s, starting with the Indian UCIL methyl isocyanate (pesticide) disaster in Bhopal in 1984, the Swiss Sandoz chemical spill in Schweizerhalle in 1986, the Ukrainian nuclear meltdown in Chernobyl in 1986, and the US PEPCON ammonium perchlorate disaster in 1988, among others. Previously, we were asked "why can't you scientists leave things alone" in *The Man in the White Suit* (1951) and "our scientific magicians" were blamed for the dystopian future in *Soylent Green* (1977). Industrial accidents were now tarnishing the public image of chemistry. They also gave the 'research-warning' and 'research-phobia' themes an eerie sense of foreboding.

A few years after *The China Syndrome* (1979), radiation is also central to *Silkwood* (1983), based on the true story of staff at the Kerr-McGee nuclear fuel rod production facility. With Meryl Streep in the Oscar-nominated titular role, technicians are shown working with glove boxes as they purify "mixed plutonium and uranium oxide" into fuel pellets. Occasionally we also see some lab glassware items like conical flasks, pipettes, and

Onscreen Chemistry: The Portrayal of Chemical Science in Film and TV
By John O'Donoghue
© John O'Donoghue 2025
Published by the Royal Society of Chemistry, www.rsc.org

beakers. Several lax work practices are shown in relation to radiation exposure, with the staff joking that the safety drills are not carried out because they "might stop production for 10 minutes". During a decontamination clean, a technician is exposed "externally" but not "internally" to plutonium with 24 dpm (disintegrations per minute), resulting in a violent scrub down.

Interestingly, Silkwood recalls that her mother discouraged her from science class in high school because "there were no girls", telling her to take home economics instead. In the movie, she gets transferred to metallography where she observes a co-worker manipulating X-ray negatives to hide the defects in fuel rods. This, along with the lax safety protocols, prompts her to get involved with the union where she learns that there is no acceptable level of exposure because "plutonium gives you cancer, flat out". She informs the Atomic Energy Commission about the lax work practises, but her co-workers start to alienate her due to fear of losing their jobs. There are also several references to missing plutonium, with reports saying that plutonium is missing from "just about every nuclear plant".

Later, she becomes contaminated internally and goes to Los Alamos for testing, where they find americium in her lungs. The doctor tells her that "americium is produced when plutonium disintegrates" which they say is equivalent to "6 nanocuries of plutonium". This is less than the "maximum permissible body burden" limit of 40 nanocuries (0.65 micrograms). However, it is also stated that although the instruments are "very sophisticated" in Los Alamos, their accuracy may be off by $+/-$ 300% at that level, but that still puts her under the limit. The movie finishes with Silkwood involved in a fatal car crash on the way to meet a journalist to expose the work practises at the facility. The facility was shut down a year later.

Sticking with the concept of side effects, we also have *Bell Diamond* (1986), about a Vietnam veteran whose wife leaves him due to a breakdown in communication and because she is unable to have a child owing to his sterility from exposure to Agent Orange. However, there is no evidence or direct link provided between the herbicide and sterility, in the movie or in the real world. Agent Orange is a real herbicide that was used during the Vietnam War to clear jungle growth, reduce possible ambushes and cause food shortages. However, it also contained trace amounts of dioxin, which is known to cause a variety of health effects and will be discussed again later in relation to *No Time to Die* (2021).

Finally for this section, we return to nuclear chemistry with *Fat Man and Little Boy* (1989), based on the true story about the development of the first atomic bombs, later retold in *Oppenheimer* (2023). Like the 2023 version of this story, it revolves around the relationship between General Leslie Groves (Paul Newman) and Robert J. Oppenheimer (Dwight Schultz). It also explores Oppy's internal conflict about making 'the bomb', his communist party links and his mistress. However, although the chemistry references are very brief in *Fat Man and Little Boy* (1989), many of the scenes reflect the real footage featured in *The Atomic Café* (1982) documentary released only a few years earlier.

One of the key chemistry scenes provides some background to their experiments as they "separate uranium-235, then arrange for two portions of the element to be brought together suddenly. The resulting mass undergoes a spontaneous self-generating reaction". However, the team need to overcome the issue of the heavy gun barrel needed to fire a slug of subcritical material. Later, in conversation with the fictional scientist Merriman (John Cusack), an orange is used as an analogy to demonstrate "compression of a plutonium sphere to cause an implosion", which should result in a chain reaction. There is also a scene featuring a chalkboard with diagrams and equations based around velocity related to the problem of bringing two masses of uranium together.

During testing, Merriman is exposed to a large dose of radiation when a hemisphere of beryllium slips while testing a sample of glowing blue plutonium. To prevent it from going critical he touches it with his bare hand, referencing a real accident which occurred in Los Alamos a few months after the two bombs were dropped on Japan. The dazzling blue light is also true, which occurs due to Cherenkov radiation, caused by photons travelling as waves with high frequencies and short wavelengths. In the movie, Merriman suffers a slow and gruesome death from the radiation exposure, which is played out as a metaphor for what the bomb will do to its victims. Finally, worth noting is when Oppy's wife declares "what about love and understanding? I thought that's what science was all about". The movie closes by stating that Oppenheimer died of cancer in 1967.

8.2 CHEMISTRY EDUCATION

The link between chemistry and education takes a significant turn mid-way through the 1970s, now giving us more realistic depictions compared to previous versions. First, it may come as a surprise, but there is a brief scene in a school lab in *Grease* (1978). In the background is a skeleton, a shelf with chemical containers, some beakers, and a few preserved animal specimens. However, the only activity that takes place is a prank involving a boy placing a preserved frog in a girl's handbag.

We then cross the pond to a Scottish school lab in the romantic comedy *Gregory's Girl* (1980), which follows a teenage boy who falls in love with a classmate when she joins the football team as the only girl. There is a brief scene in the school lab with the two female protagonists chatting in front of beakers. The blue solution is on a magnetic stirrer and the green solution contains a pH probe. Dorothy asks her classmate to "pass the sulfuric acid", who then asks "what's the pH in that?", which is stated as "seven". Dorothy uses a small pipette to add a few drops of acid to her beaker and the scene ends when she asks for "the bromide". There is also an accurate poster in the background showing the process of catalytic cracking used in refineries to break down large hydrocarbon molecules into smaller ones.

Although the scene is very brief, it contains lab skills like using a pipette and realistic instruments like a microscope, pH meter and magnetic stirrer. There are no random tubes joining glassware together and other than the brightly coloured solutions, it is a very reasonable depiction of a school lab experiment. However, unfortunately, there is no mention of what the experiment is, they aren't wearing safety goggles, and they aren't wearing lab coats. Later, Gregory is chatting to a friend while developing photos in a darkroom. His friend pours something from one measuring cylinder into another as they speak and then proceeds to drink it, which is of course not allowed in any lab. His friend also complains about the cost of developing photos, stating that the "chemicals, and the paper, are really expensive".

In the same year we also have the French language comedy *Les Sous-doués* (1980), about a group of students at a private school who play pranks on their teachers. Eventually one of their pranks backfires when their fake bomb gets mixed up with a real terrorist attack. When making the fake bomb, they use a children's chemistry set, a clock, and some shotgun cartridges. Worth a brief mention is another example of movies predicting the future, when a student uses a device to translate French to English by typing it. He says that one day he will be able to speak to it, like modern smartphone apps.

There is also a scene in the chemistry lab where the teacher explains that they need to use a pipette to add a (pink-coloured) catalyst into a beaker of crystals. This produces smoke and the teacher explains that smoke means there's a reaction. One of the students pours an entire flask of the pink catalyst into the crystals, which produces a large cloud of smoke (most likely dry ice – CO_2). The teacher ignores the prank and writes $Cl^2HC–CHCl^2$ on the board, which is incorrect for trichloroethane as he says. He also uses superscript instead of subscript for the number of chlorides!

We then graduate from the school lab to the college lab in the US comedy *Real Genius* (1985), which follows a new student who is assigned to work with a senior student (Val Kilmer) on a "5 megawatt chemical laser". However, they later discover that their corrupt research supervisor is working for the military, who plans to use their laser as a space-based weapon. The movie opens in a high-school science fair where the new student is showcasing his "flashlamp-pumped ultraviolet laser". He describes it as an "atomic iodine laser" with a wavelength of 342 nm. Although this is a real-world gas laser that was first described in the 1970s, it operates in the near-infrared (1315 nm), not the ultraviolet (342 nm).

Most of the movie features similar antics to those seen in *Animal House* (1978), but with a distinct science theme. For example, a student is annoyed that her sleigh didn't go as far as she wanted on the makeshift ice slope due to its "drag coefficient". There are also common tropes depicted among the science students such as clumsiness and social awkwardness. In addition to this, the lasers usually start a fire or cause something to explode when tested, in keeping with the 'chemistry causes explosions' trope. There is also an example of someone eating from lab glassware and students sabotaging experiments due to their competitive nature. However, there is also a brief positive discussion about the need for work/life balance, because "all science and no philosophy" can be harmful.

Later, while working on a cyanide-based laser, they decide to use liquid nitrogen, claiming that "it is possible to synthesize excited bromide in an argon matrix – yes, it's an excimer frozen in its excited state". This will produce "one kilojoule per cubic centimetre at 600 nm, or 1 megajoule per litre". However, the colour of the laser beam is shown as purple (ultraviolet), as it should be for an excimer-type laser, but 600 nm implies that the colour would be orange. Although there is a scientific basis for the type of laser described here, the power output of 6 megawatts doesn't even come close to the most powerful lasers in the world today which are measured in petawatts! We will return to the use of liquid nitrogen onscreen later.

Finally, we have the Oscar-winning drama *Dead Poets Society* (1989), about an English subject teacher (Robin Williams) who instils a love of literature, poetry and acting in his students at a private boarding school. The movie frequently puts these 'creative pursuits' in contrast to "medicine, law, business, engineering", which are described as "noble pursuits" and "necessary to sustain life". However, "poetry, beauty, romance, love, these are what we stay alive for". The movie opens with the oppression of 'creative pursuits' in favour of 'noble pursuits', when a student (Neil) is told by his overbearing parents that he needs to drop extra-curricular activities to focus on preparing for medical school. These nonsensical attitudes to the 'creative pursuits' force the students to play music, act, and recite poetry at the titular secret society, away from the school buildings.

Compared to Robin Williams' character, all the other teachers are portrayed as strict and conservative. Before we are introduced to the English class, there are brief scenes showing other subjects like Latin and Maths. The chemistry lab also appears with some glassware. The teacher tells the students to "pick three laboratory experiments from the project list and report on them every 5 weeks". He also hands out a red chemistry textbook and tells them "the first 20 questions at the end of chapter one are due tomorrow". The distinctive red chemistry book appears again later when Todd is struggling to write a poem, scrunching up the paper and opening the chemistry textbook instead. It also appears near the end, positioned under the *Five Centuries of Verse* poetry book, representing the fact that Neil has chosen arts over science, against his father's wishes.

Chemistry is a requirement for medicine courses at most universities around the world, so the red chemistry textbook is used a metaphor for Neil's struggle between his desire to pursue acting and his father's desire for him to pursue medicine. However, the book is also an example of anachronism.[†] Although the movie is set in 1959, the red chemistry textbook is titled *Chemistry: A Modern Course* by Smoot, Price and Smith, which is copyrighted 1987.[1] Overall, *Dead Poets Society* (1989) reinforces the trope of arts *vs.* science, ignoring the creativity needed for scientific innovation and the beauty of natural phenomena.

8.3 SPACE CHEMISTRY

As previously mentioned, the concept of sleeping for hundreds of years and waking up in an unrecognisable future first appeared in the 1910 H. G. Wells novel *The Sleeper Awakes*. In the real world, slow programmable freezing was first developed during the early 1970s, which eventually resulted in the first frozen human embryo birth in 1984. These advancements, and the widespread availability of cryogenic liquids like nitrogen and helium, fuelled new storylines and onscreen depictions, like *Real Genius* (1985) as mentioned already. It also sparked a new wave of fascination with cryogenics. After previously meeting Buck Rodgers as a 12-part film serial in 1939, we return for more adventures in *Buck Rodgers in the 25th Century* (1979–1981). This time our protagonist returns to a dystopian earth after 500 years frozen in space. However, in contrast to the 1930s version, reanimation from cryostasis is described here as "difficult and precise".

[†] An anachronism is a chronological inconsistency, usually in relation to people, events, objects, language terms and customs from different time periods. As well as the chemistry textbook, there are several anachronisms in *Dead Poets Society* (1989) which is supposedly set in 1959. Other examples include a desk packaged in shrink wrap, which wasn't invented until the 1970s, and the tune played by the bagpiper is *The Fields of Athenry*. Although this tune is often considered a classic Irish folk tune, it was only written in 1979.

Unlike the fictional Nirvano gas mentioned in the 1939 serial, this time Buck "was frozen by a combination of gases like oxygen, cryogen, ozone, methylon, almost in perfect balance". Oxygen and ozone are of course real gases, but 'cryogen' is a term used to describe any substance for obtaining low temperatures, *e.g.* a refrigerant. However, the closest in name to methylon is methylone, which is a real-world amphetamine substitute for MDMA. There is also a chemistry reference in relation to a burn mark on his ship, when he says "those burns are fresh, the cordite is still unoxidized". Cordite is a propellent made from a blend of nitrocellulose, nitroglycerin and petroleum jelly. However, it is unlikely to be used in a weapon in space, and it wouldn't oxidise in space so it would be difficult to age the markings based on that concept.

Overall, the movie and TV series follows Buck's adventures as he brings traditional methods to a future which has embraced science to such an extent that they've lost their "gut feeling" and "awareness". This concept is used again in *Demolition Man* (1993) when Sylvester Stallone's character is awoken from a "CyroPrison" in the future to fight a criminal from his era. Here, future society has evolved beyond violence, so the future police force has no experience in stopping violent criminals and needs a "caveman" to show them how. Like the 1980s Buck Rodgers, *Demolition Man* (1993) portrays reanimation as a complex medical process. Worth a brief mention here as well is that *Demolition Man* (1993) pays homage to *Metropolis* (1927). Both movies feature an idyllic life in a futuristic city above ground while others toil away in poverty underground. References to *The Sleeper Awakes* and *Metropolis* (1927) will appear again later in the animated series *Futurama* (1999–present).[‡]

Next, we have the three-part sci-fi TV mini-series *V: The Final Battle* (1984), which follows a group of resistance fighters in a guerilla war against extraterrestrial visitors. Several references and brief scenes in labs and medical facilities are shown

---

[‡] The concept of 'evil' people living underground, and 'innocent' people living above ground probably traces its origins to the 1895 H. G. Wells novel *The Time Machine*. In *The Time Machine* the subterranean people are called 'Morlocks', and their origins as working class people are described in *The Sleeper Awakes*. It is also theorized that the Morlocks are a reference to neanderthals living in caves since the first fossils were discovered in the mid 19th century.

throughout as they create a super-weapon, understand the visitors and treat wounded soldiers. However, the most relevant scene takes place in the third part when they are comparing bacteria through a microscope in a glove box. In the background there's an accurate-looking periodic table and a bench with test-tubes, sample vials and glassware filled with a red solution, mostly connected with tubing. Eventually they discover a bacterium that the visitors have no defence against, referencing a similar concept used in *The War of the Worlds* (1953).

Finally, we then return to the concept of evil artificial intelligence (AI) in the British sci-fi thriller *Saturn 3* (1980). The plot revolves around a rogue scientist (Harvey Keitel) who kills and impersonates a cargo ship captain to gain access to the "experimental food research station". The station is located on Saturn's third moon and is run by two other scientists (Kirk Douglas and Farah Fawcett). Almost immediately we are introduced to drug use in the form of "blue dreamers", described as a sleeping pill, while another pill is later described as a "3D inner experience". The rouge scientist builds a robot with a 'brain', filling it with colourful liquids run through lab condensers and tubing for no apparent reason other than for visual effect.

The robot teaches itself through a type of neural link with the rogue scientist, described as a "brain-to-brain" connection. However, from these tutorials the robot learns about the murder, which turns the AI robot into a killer as well. In a nod to Frankenstein's monster, the AI robot eventually turns on its creator, smashing some colourful glassware in the process. Noteworthy is the fact that the AI robot needs to recharge its batteries. This is in contrast with *The Terminator* (1984), which uses a "nuclear power cell" that can last "120 years", although this isn't mentioned until *Terminator 2: Judgement Day* (1990).

8.4 ROBOTIC CHEMISTRY

Next, we stick with the theme of cyborgs and robots with the US sci-fi satire *Robocop* (1987), whose armour is described as "titanium laminated with Kevlar". Unlike the nuclear-powered Terminators and the rechargeable robot in *Saturn 3* (1980), it is never mentioned what powers Robocop. However, he is told to rest in a specific chair at the station, so perhaps this is where he recharges his batteries. Also, in contrast to most other cyborg depictions, Robocop is the hero.

We also revisit the themes of dishonest companies and rec- reational drugs when Robocop tracks down the crime boss who shot him when he was fully human. The drug is called "blow" (slang for cocaine) and is only used in one scene before showing the drug production factory. Strangely, the machinery is filling small volumetric flasks with the powder, which are then pack- aged into boxes. Volumetric flasks are used for liquids, not powders, making this is an odd and expensive way to distribute cocaine. There are also yellow tubes, chemical containers and some colourful liquids shown in the background. Robocop breaks down the door and shoots all the henchmen, crushing flasks as he walks. Later, one of the remaining henchmen cra- shes into a large container labelled "toxic waste" which dissolves his skin, but there's no explanation provided for this.

Noteworthy is a news report about a space-based laser weapon which accidentally kills over 100 people during testing, refer- encing the plethora of other movies featuring this concept dur- ing this era. The scientists in *Robocop* (1987) are always wearing white lab coats, which continues in the sequel with some subtle differences. The sequel *Robocop 2* (1990) continues the theme of drugs with a new fictional drug called "nuke", led by the crime boss Cain. It also opens with a news report about the Amazon Nuclear Power facility "blowing its stack and irradiating the world's largest rainforest". Interestingly, the evil cyborg (Robocop 2) has a radiation symbol (trefoil) on its chest, imply- ing that it is nuclear powered. This also serves as a double en- tendre in conjunction with the name of the new drug.

The nuke drug is shown as a red liquid in a micro-syringe package, administered through the skin in the neck. The first drug-processing lab onscreen contains barrels with 'highly flammable' hazard symbols, as well as conical flasks, measuring cylinders and condensers with some tubing. There are also some chemical containers on shelves and metal funnels, but every- thing transparent is filled with the bright red nuke. The staff are wearing aprons, which dissociates the scene from the pro- fessional scientists shown later in lab coats. Robocop busts the makeshift lab, but they relocate to a more industrial location.

The new lab contains shelves of chemical containers, volumetric flasks, beakers, and instrumentation. There are also large metal reaction vessels around the site, presumably for the large-scale manufacturing of nuke. Robocop again busts the operation, but the company uses the head criminal to make a new obedient cyborg by taking advantage of his nuke addiction. Interestingly, one of the company scientists in a lab coat protests when the executive uses nuke to control the new cyborg, a further example of dissociating the professional and illegal sides of chemistry. Finally, worth a brief mention is a news report that states people can only spend 20 seconds in the sunlight in California because "we lost the ozone layer". The model then proceeds to lather herself in a thick blue sunblock, but the warning label says "frequent use will cause skin cancer".

Finally, we have the British sci-fi horror *Hardware* (1990), set in a dystopian, irradiated, and overpopulated future. The story revolves around a self-repairing military 'droid' called the Mark 13. The data file for the droid states that it can "recharge its storage batteries from just about any power grid, including the sun". Later, the file also states that it carries an "additional chemical weapon, a cell destroying toxin, 3-psilocybin morphate", which it can administer by injection. Psilocybin is a real-world psychedelic compound found in magic mushrooms, and the morphate suffix is a fictionalisation of morphine. The movie states that the drug causes sensory deprivation and death, and the character says it "smells like apple pie". It also kills within seconds and even makes the victim "enjoy the experience". Other chemistry references include a character who uses a fictional drug to get high and a brief mention of radioactive iodine, which is used in the real world to treat thyroid cancer.

8.5   FICTIONAL MATERIALS

Up to now, we have seen several examples of fictional chemical materials, such as in *The Man in the White Suit* (1951) and *Monkey Business* (1952) among others. In most cases the fictional chemistry is an extrapolation or exaggeration of real-world chemistry concepts, glassware, and terminology. But there is another form of fictional chemistry where the concept is entirely invented with little basis in reality. A good example is the comic hero Superman who first appears on screen in 1941 in a series of animated shorts, followed by two movie serials in 1948 and 1950. The relevant example here is of course the fictional substance kryptonite, not to be confused with the real chemical element krypton (ironically, this is inert).

In *Superman III* (1983), Lex Luther performs a chemical analysis of kryptonite using a space laser and a supercomputer. The output displays "plutonium 15.08%, tantalum 18.06%, xenon 27.71%, promethium 24.02%, dialium 10.62%, mercury 3.94% and unknown 0.57%." Most are real chemical elements except for the unknown and dialium, which is never explained. There is also a warning on the output: "when the elements combine, an intense heat of fusion occurs". The character (Richard Pryor) switches the "unknown" to "tar", inspired by a cigarette packet, and sends the list off to the "boys at the lab" to make synthetic kryptonite. It appears a short time later as a bright green crystal, delivered on a platter by a scientist in a lab coat as if it was food prepared in a kitchen. There is another reference to synthetic substances later, when Superman crushes a lump of coal in his hand to make a diamond.

There is also an interesting scene in relation to industrial accidents in which Superman stops a fire at a "chemical plant". All the workers are shown wearing white lab coats and there is a brief look inside a lab with some glassware filled with colourful solutions and connected by tubing. The selfless chemist refuses to leave because he needs to watch the (fictional) "concentrated beltric acid. If that stuff heats up over 180 degrees, we've got a crisis on our hands". He goes on to say that "if it begins to heat up, it'll turn volatile. If that happens, you'll get a great cloud of smoke that'll eat through anything – steel, concrete, anything.". Thankfully, Superman saves the day by freezing a lake and dropping it over to the plant to create rain.

Superman III (1983) uses real-world themes like chemical-plant accidents and mixes this with fictional materials (dialium). Although there is a fine line between exaggerated and entirely fictional chemistry, the latter is usually associated with non-sensical or made-up wording and terminology. Another good example of a fictional material is the classic TV series *Knight Rider* (1982–1986), about the "self-made billionaire" Michael

Knight (David Hasselhoff) and his autonomous car KITT (Knight Industries Two Thousand).

The AI-controlled car has an armoured "tri-helical molecularly bonded shell", also called "plasteele 1000", which protects it from almost all forms of conventional firearms and explosive devices. KITT can only be harmed by heavy artillery, lasers, and rockets, and even then the blast usually leaves most of the car's shell intact. The shell also protects against fire and electricity; however, it is vulnerable to acids and prolonged exposure to seawater. It is explained that the shell is a combination of three secret substances, referred to as "The Knight Compound". At one point, Garthe Knight steals the formula and reproduces it to make his truck invincible in *Goliath Part 1 and 2* (1983) and *Speed Demons* (1984).

Total Recall (1990) also provides us with a fictional material, which is core to the plot. In the future, humans are mining a fictional mineral known as turbidium on the planet Mars for "military applications" which acts as a type of MacGuffin to move the plot along.[§] There is also a discussion about a reactor and chain reactions in relation to turbidium, so perhaps it is an energy source. Interestingly, there is also a reference to *Flash Gordan's Trip to Mars* (1938) when it is claimed that "They just want our turbinium so they can zap things from space". The name 'turbidium' is extrapolated from the real scientific term 'turbid', meaning cloudy, opaque, or thick with suspended matter. As a result, the name used for the mineral is most likely poking fun at the non-specific details given about turbidium. We will return to fictional materials again later, particularly in relation to the Marvel Cinematic Universe (MCU).

8.6 SUPERNATURAL CHEMISTRY

First here is the sci-fi horror *Altered States* (1980), about a psychopathologist who begins self-experimenting with sensory deprivation and hallucinogens using a flotation tank. It explores

---

[§] A MacGuffin is an object, event, or character in a film or story that serves to keep the plot in motion, despite usually lacking intrinsic importance. Usually, the MacGuffin is revealed in the first act and the importance of it can vary throughout. A well-known example of a MacGuffin is the briefcase in Pulp Fiction (1994), which motivates several of the characters but whose contents are never revealed. Similarly, the plot of Ronin (1998) entirely revolves around a briefcase with unknown contents.

themes of consciousness and religion coupled with a thin veil of scientific research and methods. There is a brief scene shown in a lab with some large containers, tubing, and wash-bottles, among other items. Later, there are also some images of beakers and chemical containers in relation to the hallucinogen. Like the early gothic horror depictions, the protagonist tries to convince his peers and others about his theories in relation to consciousness. However, his request to repeat the self-experimentation in the interest of "scientific validation" heavily hints at addiction. His partner also returns us to the theme of 'mad scientists' when she states "I think he's on the verge of a breakdown, he was here all afternoon and carried on like a madman".

Next, we have the Canadian sci-fi horror *Scanners* (1981) about superpowered individuals capable of telepathy and psychokinesis called 'scanners'. The franchise centres around a drug called "ephemerol", which "restores sanity" by suppressing scanning abilities. However, one of the characters believes that the scanners may be the "next stage in human evolution". It is later revealed that he founded a pharmaceutical company (Biocarbon Amalgamate) to make the drug for pregnant women to turn their children into scanners. The drug was originally developed as a sedative, but when they discovered the side effect, they decided to use it to create a "generation of scanners to take over the world".

This is a reference to the real-world thalidomide scandal, in which thousands of children were born with severe deformities in the 1950s and 1960s. Thalidomide was first developed as a tranquilizer and later acquired by a Germany-based pharmaceutical company established by a former Nazi party member. Development was also headed by a former Nazi chemist who escaped prosecution for his experiments on prisoners in concentration camps related to creating a 'superior Aryan race'.[2] Despite knowing about the severe side effects and history of the drug, thalidomide was sold in pharmacies in Canada until 1962.[3]

The name of the fictional drug 'ephemerol' may be a play on the real word 'ephemeral', which means 'lasting for a very short time'. The suffix '-ol' implies that the drug is an alcohol-based compound, *i.e.* organic compounds that carry at least one hydroxyl functional group bound to a saturated carbon atom like methanol, ethanol and cholesterol. Two sequels and two spin-offs followed, using similar concepts in addition to themes of drugs and addiction. In *Scanners II: The New Order* (1991) a variant of ephemerol (Eph2) is found to be addictive and causes physical damage over time. It includes a scene in a lab where the test subjects are freed, and their addiction is later treated. Finally, in *Scanners III: The Takeover* (1992) the third variant of the drug (EPH-3) is a patch like real-world nicotine patches, but it causes the protagonist to become unhinged and commit murders. The rampage of the protagonist ends when the EPH-3 patch is removed, so the madness is entirely due to a side effect of the drug and not the person.

We now move from the big screen to the small screen for *The Incredible Hulk* (1977–1990) TV series and movies, which follows Dr Banner, a physician based at a medical research institute. The TV movie pilot (1977) presents him in a lab surrounded by conical flasks, beakers, test-tubes, reflux condensers, and a periodic table of elements. He explains that he is researching people who demonstrate superhuman strength. Later, while examining mitochondria with an electron microscope, they increase the magnification from 200 000 to 1 million to look at the DNA structure. On the computer screen adenine, thymine, guanine, and cytosine are listed, as well as some real (H_2O, CH_4) and some impossible (CH_4NH_4) chemical structures. The structures shown onscreen are also moving, which would not happen in an electron microscope.

Later, Dr Banner makes a connection between gamma ray radiation from solar flares and a DNA abnormality. He decides to self-experiment using an X-ray machine, switching the dial to "gamma". Although X-rays and gamma rays have the same basic properties, they come from different parts of the atom. So, it is unlikely that an X-ray machine would also produce gamma rays on demand. Later, in a fit of rage he turns into the green Hulk (Lou Ferrigno), returning us to the story of Dr Jekyll and Mr Hyde once again. However, unlike the classic gothic horror story, the Hulk (Mr Hyde) uses his super-strength to help people. Also relevant to this discussion is when his colleague is hurt by a fire in a room labelled "Chemical Storage" that explodes and destroys the institute. This once again emphasizes the connection between the word 'chemical' and explosions. The series finishes with the TV movie *The Death of the Incredible Hulk* (1990), in which Dr Banner tries to rid of himself of his alter-ego with help from another scientist featuring several scenes in a lab talking about DNA bases.

Next, we return to the theme of chemical side effects in a homage to *The Incredible Shrinking Man* (1957). In the original movie, the shrinking was caused by a combination of insecticide and radiation. There is no reference to radiation this time; instead, *The Incredible Shrinking Woman* (1981) concentrates entirely on chemicals with a combination of household items and exhaust fumes causing the protagonist to shrink slowly over time. Interestingly, the movie pokes fun at the rapid expansion of new products in the 1980s and includes evil corporations who try to cover up the side effect when the protagonist tries to warn others.

While at a medical institute for "unexplained phenomena", she undergoes "a radioactive iodine test and a scanning electron microscope test". Later, a secret collaboration of scientific companies attempts to capitalise on the side effect by extracting a serum from her blood to "shrink entire nations" to their advantage. This concept is partially explored again later in *Downsizing* (2017). Eventually, the protagonist of *The Incredible Shrinking Woman* (1981) is kidnapped and locked in a cage in a lab, thus providing us with the unusual visual image of, from her perspective, giant glassware. This concept will be seen again later in relation to the animated TV series *Pinky and the Brain* (1995–1998). She escapes the cage but is then held in place using a glass funnel. Unlike the original movie, the shrinking is eventually reversed by another accidental combination of consumer products.

Noteworthy is the difference between *The Incredible Shrinking...* movies and the better-known *Honey I Shrunk the Kids* (1989) franchise. In the latter example, although the shrinking is still accidental, it occurs due to a purpose-specific device, *i.e.* the 'shrink machine'. In both of *The Incredible Shrinking...* movies, a combination of chemicals and/or radiation is what causes the unexpected side effect. In the real world, the combination of certain products can be harmful, such as bleach and ammonia which produces toxic gases called chloramines.

We then move onto the US horror *The Serpent and the Rainbow* (1988), directed by Wes Craven, about a pharmaceutical company who wants to obtain the formula to a Haitian 'zombie drug' to use as a new super-anaesthetic. It is loosely based on the 1985 non-fiction book of the same name by Wade Davis in which he recounts his experience in Haiti. Narration is provided by a consultant from the company (Bill Pullman) who travels to Haiti to find the source of the drug and bring it back to Boston. It starts with some scenes in medical clinics and mentions of real-world drugs like Thorazine (chlorpromazine), an antipsychotic. Eventually he is given the yellow powder, which he brings back to the US where they run tests on baboons. However, most of the movie is about the psychosis associated with being buried alive while in the 'zombie' state.

The movie finishes with a note saying the active ingredient of the zombie powder is tetrodotoxin (Figure 8.1), a real neurotoxin associated with the pufferfish. Tetrodotoxin is a popular plot device for fake deaths onscreen, appearing in *Hello Again* (1987), *The A-Team* (2010), *Captain America: The Winter Soldier* (2014), and *War* (2019), among others. It is also used for fake deaths in several TV series such as *Miami Vice* (1984–1990), *CSI: NY* (2004–2013), *Chuck* (2008–2012), *Nikita* (2010–2013) and *Jane the Virgin* (2015–2019). In *MacGyver* (1985–1992), a *Datura stramonium* leaf is described as an antidote, but there is no evidence for this since it has its own toxicity. The paralysis caused by tetrodotoxin is also used for torture in *Law Abiding Citizen* (2009) and *Alex Cross* (2012), and as a murder device in *Columbo*

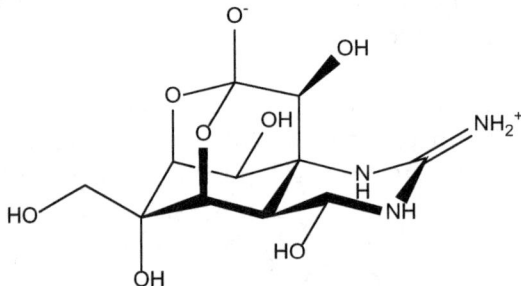

Figure 8.1   The chemical structure of the highly potent real-world neurotoxin tetrodotoxin. It is a very popular plot device for fake deaths, torture and poison in movies and TV.

(1971–2023), *Covert Affairs* (2010–2014) and *Archer* (2009–present). References to a pufferfish/blowfish containing a toxin also appear in *The Simpsons* (1989–present), but the name of the toxin is not mentioned.

Finally, we finish this section with the Mexican–Italian *Santa Sangre* (1989), an avant-garde psychological horror about a boy called Fenix who grows up in the circus and struggles with his childhood trauma. As an adult in a 'mental asylum', he recalls the trauma in a brief scene in which his mother pours sulfuric acid (H_2SO_4 is shown on the label) over his father for cheating on her. However, it is not explained why the sulfuric acid is stocked in the circus. There is also a brief scene in a makeshift lab as Fenix tries to replicate *The Invisible Man* (1933) while watching the movie on TV. He mixes some colourful solutions in an elaborate glassware setup and drinks from a measuring cylinder, but he is unsuccessful in turning invisible.[¶]

8.7 KILLER CHEMISTRY

We start here with the French thriller *Eaux Profondes* (1981), which follows a perfumier who kills the lovers that his wife brings home. We only get a few brief scenes in the perfume lab which contains several shelves stocked with hundreds of liquid samples in a variety of colours. There are also some large containers on the bench, test-tubes, a condenser set up for reflux attached to a retort stand, as well as some mechanical devices. The perfumier provides the distributor with various samples from volumetric flasks, one of which is sealed with parafilm. Parafilm is a distinct semi-transparent sealing film used in labs made from waxes and polyolefins. It was invented in 1936 and this is probably the first obvious appearance onscreen.

We stay in France next for *Mauvais Sang* (*Bad Blood*) (1986) about a new sexually transmitted disease (STD) which only affects young people. A 'serum' is available, but it is locked away in a pharmaceutical company who refuse to release it. The movie centres around two aging crooks and their apprentice who are

---

[¶] A word of warning that *Santa Sangre* (1989), called *Holy Blood* in English, contains sexual violence, domestic violence, depictions of suicide, nudity, and gore. The writer Roberto Leoni worked in the library of a psychiatric hospital where he had been in contact with people suffering from mental disorders such as dissociative identity disorder.

blackmailed into to stealing the serum for financial gain. In a brief scene, while stealing the serum, a lab is shown with a bench of glassware joined together with tubing. There are also interesting references to Halley's Comet throughout, which is blamed for extreme weather changes like a heatwave and sudden snow. The comet was visible from Earth in 1986 and previously mentioned here in relation to the plot of *The End of the World* (1916). The next time it will be visible is 2061.

Next, we have some 14th century forensics in *The Name of the Rose* (1986), about a Franciscan monk (Sean Connery) called William Baskerville who investigates a series of mysterious deaths. His name is derived from the natural philosopher William of Ockham and the well-known Sherlock Holmes story *The Hound of the Baskervilles*. Echoing Sherlock Holmes, he demonstrates keen observational skills throughout and even uses the phrase "it's elementary". As William investigates the corpse of a victim, he is assisted by an apothecary whose workshop is filled with boiling liquids, wooden apparatus, and ceramic pots. When asked what he uses arsenic for, the apothecary replies that "it is a most effective remedy for nervous disorders, if taken as a compound in small doses". The victims all have one thing in common, black on their fingers and a black tongue.

While investigating the writing desk of the victim, William uses a candle flame to reveal hidden Greek writing on some paper, telling his assistant that it was written with lemon juice. This is accurate since heating dried lemon juice causes some of the sugars to react with oxygen in the air and turn brown. This contrasts with *National Treasure* (2004), in which they incorrectly use lemon juice to reveal a hidden message on the back of the US Declaration of Independence. Later, a monk is shown stealing a large container from the apothecary workshop, which turns out to be lime leaves. The apothecary tells William that lime leaves in a bath are used to alleviate pain. This is true for Kaffir limes (also call Thai lime), which are native to Southeast Asia. Finally, we discover that the pages of a forbidden book have been poisoned, presumably with arsenic, so when a reader touches it and then touches their mouth, it kills them. In the real world, some reports of arsenic poisoning from the 19th century includes references to a dry and black tongue.[4]

Finally, we have a combination of chemophobia, evil industry and toxicology in Tim Burton's *Batman* (1989). We start with Jack/Joker (Jack Nicholson) attempting industrial espionage in Axis Chemicals, where he complains about the fumes from the containers labelled "Toxic Gas" and "Highly Toxic". During a fight with Batman, Jack falls into a vat of green sludge which disfigures him into the Joker. Later he uses an old CIA file labelled "DDID nerve gas – Discontinued 1977" to create Smylex, a fictional nerve agent that he laces into hygiene products. This causes fatal convulsions of laughter, with two runway models reported as having died from a "violent allergic reaction, although drug use has not been ruled out".

The police dossier on Jack/Joker claims that his aptitudes are in "science, chemistry and art", with 'chemistry' standing out to Batman. This prompts Batman (Michael Keaton) to collect products for analysis in the Batcave. Similar to *The Incredible Shrinking Woman* (1981), he initially claims that the mixing of consumer products is what caused the toxicity to build up in the victims. On the computer screen there are random words like "product, stable, 39% water, nucleus, wt 1.03 g". Finally, the Joker lures people to a parade with the promise of free money where he plans to poison everyone using Smylex gas from the parade balloons. Batman later removes the balloons and saves the day. Interestingly, there is a scene where Bruce attempts to explain that he is Batman by saying "you know the way people have different sides to their personality. Sometimes a person will actually have to lead a different life", echoing the Dr Jekyll and Mr Hyde story once again.

8.8 THE LAST LAUGH

We finish this chapter on a somewhat lighter note with three comedies featuring some onscreen chemistry. First is the US satire *Polyester* (1981), which pokes fun at suburban life. It opens in a lab with colourful solutions in flasks and a scientist looking through a microscope. He explains "odorama" which allows movie audiences to smell fragrances linked with the scene. He explains that he has developed the scratch-and-sniff card after being locked away in the lab for many years. Like other comic depictions, the scientist has a European accent – in this case it is somewhat German. There is also a scene where the teenager uses a cloth to sniff some of the household cleaners under the sink.

Next is the British medical comedy *Britannia Hospital* (1982), which satirises the 'mad scientist' concept with references to *Frankenstein* (1931) and *The Mad Ghoul* (1943), among others. The story follows a journalist trying to expose the professor who is carrying out human experimentation. It features only one scene in a lab as the professor blends a human brain for consumption from a beaker. He is surrounded by conical flasks, volumetric flasks and beakers containing colourful liquids dotted randomly around the room among microscopes. Like Dr Jekyll, there is also a scene where he defends his legitimacy but is called mad by his peers.

Finally, we have *Top Secret!* (1984), a parody of Elvis musicals and spy movies in which an American rock-and-roll singer becomes involved in a resistance plot to rescue an imprisoned scientist. The scientist is shown in a "secret prison lab" which contains test-tubes, measuring cylinders, volumetric flasks, and miscellaneous electronic equipment. He explains that before he was captured he was developing the first "magnetic desalination process, capable of removing the salt from over 500 million gallons of seawater a day". For context, this is equivalent to 757 Olympic-sized swimming pools. He then goes on to explain that they are forcing him to create a "horrible weapon", called the "Polaris mine" which features a super-magnet. Noteworthy is a brief image of a fictional erotic ACME Lab Equipment calendar.

REFERENCES

1. R. Price, J. Smith and R. Smoot, *Chemistry: A Modern Course*, Merril Publishing Co., 1987.
2. R. Williams, The Nazis and Thalidomide: The Worst Drug Scandal of All Time, *Newsweek*, https://www.newsweek.com/nazis-and-thalidomide-worst-drug-scandal-all-time-64655 (accessed 2024-02-10).
3. J. F. Webb, Canadian Thalidomide Experience, *Can. Med. Assoc. J.*, 1963, **89**, 987–992.
4. W. Tomkins, Poisoning by Arsenic – Coroner's Inquest, *Prov. Med. J. Retrosp. Med. Sci.*, 1843, **6**(141), 215–218.

SEASON 9

Conspiracy Chemistry

9.1 CONSPIRACY CHEMISTRY

Conspiracy theories and coverups have been around since at least Roman times, most famously associated with the emperor Nero who is said to have faked his own death.[1] However, the secrecy of the Cold War generated plenty of new theories, giving us famous movies like *The Manchurian Candidate* (1962). Real-world coverups like the Watergate scandal in the US also sowed the seeds of suspicion, a theme which was explored around the same time in *The Conversation* (1974). The fascination with conspiracy theories and coverups accelerated after the fall of the Berlin Wall in 1989, which brought to light many real secrets as shown in *The Lives of Others* (2006). Away from chemistry, other well-known movie depictions include *JFK* (1991), about the Kennedy assassination, *Conspiracy Theory* (1997), which highlights several theories, and *Enemy of the State* (1998), about surveillance.

The conspiracies usually involve a government coverup, as opposed to the industry coverups discussed previously. An early sci-fi example of this is *They Live* (1988), about sunglasses which uncover the 'real world', a concept reimagined later in *The Matrix* (1999). First here in terms of onscreen chemistry is the French mystery *Agent Trouble* (1987), about a journalist attempting to

Onscreen Chemistry: The Portrayal of Chemical Science in Film and TV
By John O'Donoghue
© John O'Donoghue 2025
Published by the Royal Society of Chemistry, www.rsc.org

uncover a mysterious government project where 50 tourists on a bus were killed by an unknown poison gas. Near the end, a lab is shown filled with scientists in the background labelling samples and inspecting green liquids in measuring cylinders. There are also some test-tubes, syringes, and lots of white mice in cages. It is claimed that the deaths of the tourists occurred due to an accidental contamination caused by the research institute, but it later transpires that it was a purposeful test of a new nerve agent.

One of the most famous depictions of conspiracies is *The X-Files* (1993–2002, 2016–2018) franchise, whose popularity led to two spin-offs called *Millennium* (1996–1999) and *The Lone Gunman* (2001). Famously, the format of most episodes of *The X-Files* involves Mulder's (David Duchovny) desire to investigate a supernatural occurrence with the sceptical Scully (Gillian Anderson) insisting that there must be a scientific explanation. Despite Scully's efforts at a scientific explanation, she usually witnesses something she cannot explain and retains no evidence of what happened. In terms of chemistry, it mostly extrapolates or exaggerates real-world chemistry, but occasionally it also makes the leap into fictional concepts as well.

Noteworthy is an episode called *The Erlenmeyer Flask* (1993), which establishes most of the alien mythology story arc including the government conspiracy to hide the truth. An Erlenmeyer flask is a standard item of chemistry glassware, also called a conical flask or titration flask. The plot centres around Mulder accidentally discovering the titular flask labelled "purity control". After analysing the sample in a lab featuring glassware and microscopes, the guest scientist declares that it contains extraterrestrial bacteria. This is coupled with the correct chemical structure of a tripeptide shown onscreen. Later, toxic alien blood turns into a type of nerve gas when the skin of the alien–human hybrid is punctured with a needle, killing the paramedics. Near the end, Scully also discovers an alien foetus stored in liquid nitrogen, reinforcing the onscreen fascination with cryogenic liquids during this era. The main criticism here is that conical flasks are designed for swirling samples or crystallisation, not long-term storage of bacteria, since they are difficult to seal.

Although most of the science is biological or biochemical (human and alien hybrids), many of the themes discussed previously also appear throughout the franchise. Noteworthy episodes include *Die Hand Die Verletzt* (1995), featuring an evil substitute biology teacher with scenes in a school lab and hallucinogenic drugs. There is also a throwback to a classic *Star Trek* episode discussed earlier when the agents discover a silicon-based organism in *Firewalker* (1994). In that episode, Scully identifies "silicon dioxide" (sand) in a victim's lungs. There is also a throwback to side effects from chemical insecticides in the episode *Blood* (1994) and side effects from water pollution in *Quagmire* (1996). The theme of 'evil pharmaceutical companies' also appears in *F. Emasculata* (1995) and there is also a modern take on the Frankenstein story in *The Post-Modern Prometheus* (1997).

The concept of secret government labs (Hawkins National Laboratory), coverups and unethical experiments also appears again later in the recent sci-fi horror series *Stranger Things* (2016–present). However, other than some medical imagery and references to drugs which are used to alter pregnancies, the onscreen chemistry is scarce.

9.2   COLD CHEMISTRY

As discussed, the expansion of cryogenics and the widespread availability of cryogenic liquids like nitrogen and helium fuelled new storylines and onscreen depictions from the 1970s onwards. Although *The Terminator* (1984) doesn't have any real onscreen chemistry, the sequel *Terminator 2: Judgement Day* (1991) concludes with a conveniently located tanker truck labelled "liquid nitrogen". This provides the iconic image of the evil T-1000 terminator walking through the liquid nitrogen until his feet snap off. Once frozen, the good T-800 Terminator (Arnold Schwarzenegger) shoots the T-1000 (Robert Patrick), smashing him into thousands of pieces. However, the pieces then defrost and re-assemble due to the heat of the steel works around them. The movie concludes with a battle of good *vs.* evil, played out against a backdrop of cold (blue) *vs.* hot (orange) visuals.

Also noteworthy in *Terminator 2: Judgement Day* (1991) is a scene in which the protagonist Sarah Conor exclaims "men like you built the hydrogen bomb", once again blaming scientists for the end of the world because they can't leave things alone. There is also a famous dream sequence known as 'Sarah Connor's nightmare' which depicts a nuclear blast on 'Judgement Day'. The director, James Cameron, consulted with experts for the scene and was later praised "for the most realistic and scientifically accurate nuclear detonation scene in any media" at the time.

As further evidence that liquid nitrogen was now making a regular appearance on screen, it also appears in the famous 'toilet bomb' scene in *Lethal Weapon 2* (1989). The police explosives team explain that the liquid nitrogen will slow down the detonation, giving the two protagonists (Mel Gibson and Danny Glover) enough time to bail into the bathtub before the bomb explodes. This was later proven to be true on *MythBusters* (episode 178, *Toilet Bomb*; 2011). The *MythBusters* team found that an ambient-temperature bomb has a natural delay of 3.3 milliseconds and pouring liquid nitrogen onto the charge increased this to 15 milliseconds. However, when the liquid nitrogen was poured over both the charge and the battery powering the trigger, the delay was extended to 15 minutes. So, Riggs and Murtagh in *Lethal Weapon 2* (1989) could have just stood up and walked out of the situation.

In the same year that Scully discovers the alien foetus stored in liquid nitrogen in *The X-Files* (1993–2002), the big screen also depicted dinosaur embryos stored in liquid nitrogen in *Jurassic Park* (1993). The embryos are stored in 1.5 mL microcentrifuge tubes in the "Cold Storage Room", later stolen and transported in a fake shaving cream can. It is also worth recalling a famous line here which is relevant to the well-established theme of calling out scientists for not considering the moral implications of their work. After the dinosaurs begin to run amok, the character played by Jeff Goldblum lambastes the owner (John Hammond) with the line "your scientists were so preoccupied with whether or not they could, they didn't stop to think if they should".

Finally, we return to the *Batman* franchise for a cold-hearted villain in *Batman and Robin* (1997), with George Clooney flying the cape this time. It opens with Mr Freeze (Arnold Schwarzenegger) and his evil ice hockey team using a freeze gun to steal diamonds. Later, we learn that Mr Freeze was originally a "Nobel Prize winner for molecular biology" (this does not exist, only medicine) researching a cure for his wife who contracted a fictional rare disease known as MacGregor's syndrome. While attempting to cryogenically freeze her to give him more time, an unspecified lab accident causes him to fall into a tank of "cryogenic solution". This mutates his body, meaning he can't live at average temperatures and needs a cryogenic suit powered by diamonds for survival. In another reference to the financial troubles of researchers, he plans to hold Gotham to ransom by freezing it so he can save his wife.

Famously, one of Mr Freeze's quips is "what killed the dinosaurs? the ice age", and Batman refers to him as 'mad'. There is also a scene in a jungle lab with a dishevelled Doctor Isley (Uma Thurman) and the quintessential 'mad scientist' Woodrue (John Glover). Their labs contain hundreds of items of glassware filled with colourful liquids, in addition to flashing lights and electrical sparks. In a nod to Frankenstein, Woodrue uses the luminescent venom from Isley's research to create the super-soldier Bane. In fact, the entire movie uses bright and striking colours in a nod to the technicolour TV series *Batman* (1966–1968).

Isley is horrified that her research is used this way and proclaims to Woodrue that when she is through with him "you won't get a job teaching high school chemistry". This must be a terrible insult because Woodrue responds by pushing Isley into her experimental setup, destroying her lab. He also pushes all her glassware over her which includes the venom. The "poisons" and "toxins" cause the ground to sink around Isley, but she re-emerges later as the villain Poison Ivy to enact her revenge. She claims that her blood was replaced with aloe, her skin with chlorophyll and her lips with venom. She also claims that she wants to protect plants, but then proceeds to burn down the lab which includes all her plants.

9.3 DANGEROUS CHEMISTRY

We now return to chemistry which kills, explodes, and burns. First up is *The Rock* (1996), starring Nicholas Cage as a chemical weapons specialist (Stanley Goodspeed) in an exaggerated but mostly real chemistry action thriller. Here the duality of chemistry is evident throughout since it can be used to kill, but a knowledge of chemistry can also save the day. Our hero chemist explains that he is "one of those fortunate people who likes my job, sir; got my first chemistry set when I was 7, blew my eyebrows off, we never saw the cat again, been into it ever since". Interestingly, a year earlier, 'chemistry set' is also used to describe a drug lab in *Bad Boys* (1995) in which the chemists are talked down to by the gang leader. In *The Rock* (1996), the plot revolves around disenchanted Navy Seals who steal chemical weapons, one of which spills open and melts the skin of a team member.

Later, the FBI laboratory is shown filled with glassware and microscopes in the background. An alarm suddenly goes off and they suspect a package contains sarin gas, a highly toxic real-world nerve agent. The main lab contains a sealed "gas chamber" along with various glassware, instruments, and storage units. Goodspeed dons an environmental suit and enters the chamber with cockroaches, like canaries in a coalmine. However, the device activities and starts to combine two liquids to create a gas which is described as "corrosive", *i.e.* it starts "eating their suits". Sarin is corrosive in the real world, so it is normally stored as two separate precursors to extend its shelf life. However, this is a good example of how 'corrosiveness' is often misinterpreted. Corrosion is often shown onscreen as nearly immediate, but it is much slower in real life. Environmental suits are also designed to protect the occupant long enough to escape the situation. During the scene, both chemists are told to inject atropine into their hearts before their suits melt. Atropine is a real anticholinergic medication used to treat certain types of nerve agent and pesticide poisonings, but it can be injected into any muscle.

The weapons stolen by the rogue Navy seals are later revealed to be "VX gas rockets", prompting the recruitment of Goodspeed. When asked about the gas he responds "liquid, failed pesticide, discovered by mistake in 1952", which is entirely true. It was first discovered while investigating organophosphate compounds as fertilizers and was on the market briefly before it was withdrawn due to its toxicity. However, the toxicity was noticed by the military leading to its development as a chemical weapon. It was banned by the UN Chemical Weapons Convention in 1993, which likely inspired the plot of *The Rock* (1996). Goodspeed describes the functionality as "a cholinesterase inhibitor, it stops the brain from sending messages down the spinal cord within 30 seconds", which is true. Death is caused by asphyxiation *via* paralysis of the diaphragm due to a disruption caused between the nervous and muscle systems. It also persists in the environment, is odourless and is more potent than sarin.

However, it does not melt skin as claimed in the movie and the liquid is amber-like in colour, not green as it is shown. There is also a brief scene later where we can see more of the FBI lab, showing condensers clamped with tubing along with conical flasks with blue liquids on a magnetic stirrer. The VX is referred to as poison on several occasions, and disarming the rockets only seems to involve removing a "string of pearls", *i.e.* glass spheres filled with green liquid. In the real world, something caustic like sodium hydroxide would easily neutralise VX gas. Upon making the leap from the lab to field work, our hero exclaims "I'm just a biochemist. Most of the time, I work in a glass jar and lead a very uneventful life". Noteworthy also is a mention of conspiracy theories in relation to Sean Connery's 'James Bond' type character who was trained by "British intelligence" and apparently knows about "the alien landing at Roswell, and the truth about the JFK assassination". VX nerve gas is used again later in the TV series *24* (2001–2014) and *Mission: Impossible – Rogue Nation* (2015).

The concept of binary chemical systems that are only dangerous when mixed also appears a year earlier in *Die Hard with a Vengeance* (1995). In this example the red and colourless liquids are said to be harmless, but when mixed they are explosive. Unfortunately, we are never provided with the chemical makeup of the liquids or the final mixture. When one of the bombs is revealed to be a fake, one of the liquids is described as "pancake syrup".

Continuing the discussion about corrosion is *Dante's Peak* (1997), about a geologist (Pierce Brosnan) studying a volcanic eruption. While rescuing an elderly lady and attempting to escape across a lake, the metal boat they are travelling in begins to corrode. The geologist states that "the volcanic activity has turned the lake into acid", as they see dead fish floating on the surface. The boat begins to fizz, causing it to slowly sink. Acid lakes do exist in the real world. They are formed by volcanic gases which bubble up from underneath, creating sulfuric and hydrochloric acids. Hydrochloric acid will react quite strongly with the aluminium metal used for boats and since the lake is also shown as steaming hot, the acids would react even faster. Therefore, this scene is accurate since the boat would have reacted quite quickly with the acid as they crossed the lake.[2]

Next, we continue the dangerous chemistry discussion with *Fight Club* (1999). Early in the story, when the Narrator's apartment explodes, the police find residues of "ammonium oxalate and potassium perchloride", which they say indicates the bomb was homemade. This may represent another example of mispronunciation of a chemical, like that seen previously in *Monkey Business* (1952) and others. He is most likely referring to potassium perchlorate, a strong oxidizer used in fireworks and used as solid rocket fuel which will be discussed in more detail later. *Fight Club* (1999) also introduces us to lye, better known as sodium hydroxide (NaOH). While wearing safety glasses (a rarity for onscreen chemistry), Brad Pitt's character (Tyler) uses the lye to burn the skin of Edward Norton (the Narrator) which is then neutralised with vinegar. This is an accurate chemical reaction and lye would cause skin burns, but, like corrosion, the swiftness of the burn would be much slower in real life.

Later in the movie Tyler states that he needs to source large quantities of glycerine using the fat from liposuction clinics, stating "when the tallows harden, you skim off a layer of glycerine. If you were to add nitric acid, you got nitroglycerin; if you were to add sodium nitrol and a dash of sawdust, you got dynamite", which is mostly true. However, the writers purposely left out some details to prevent viewers from making dynamite themselves, since sulfuric acid is also needed.[3] There are also no details provided about how to combine them or in what quantities. The movie concludes as Tyler uses his homemade dynamite to blow up credit-card companies to wipe away debt, representing an anti-hero depiction of chemistry, *i.e.* although the chemistry is used to make something explosive and dangerous, its ultimate purpose is to improve people's lives.

9.4   ENERGY

The themes of fuels and energy are also popular during this era, inspired by real world accidents such as the Nowruz oil spill in the Persian Gulf in 1983, the *Odyssey* oil tanker blowout near Nova Scotia in 1988, and the Piper Alpha oil and gas platform explosion near Scotland in 1988, among others. Five years after the real *Exxon Valdez* oil spill in 1989 near Alaska, the plot of *On Deadly Ground* (1994) revolves around a 'blowout firefighter' and an oil-rig worker who try to prevent their corrupt and greedy employer (Michael Caine) from drilling for oil on native Inuit land in Alaska. The movie concludes with the protagonist (Steven Seagal) claiming there is a coverup by the oil industries that has prevented the adoption of alternative-fuelled vehicles. He also references the dumping of lethal toxic waste and chemical damage, but no other details are provided.

As previously seen in *Soylent Green* (1973), the theme of energy is also prominent in dystopian movies. This continues in *Mad Max 2: Road Warrior* (1981), which revolves around a petrol refinery in a not-so subtle reference to the human obsession with fossil fuels, even in a dystopian wasteland. Their source of water is never mentioned, but they are all prepared to die to gain control of the petrol ("guzzolene") from the refinery. Because petrol only has a short shelf-life of a few months before it spoils, a refinery is correctly needed to make fresh batches. Following this, the fuel in *Mad Max: Beyond Thunderdome* (1985) switches to methane gas, which is extracted from pig excrement. The methane is used to power a train engine as a generator for electricity in Bartertown. But since the pigs need to be fed large quantities of food, this a hugely inefficient source of fuel.

Unlike previous movies in the franchise, water does make multiple appearances in *Mad Max: Beyond Thunderdome* (1985), initially in relation to a salesman with a tank in Bartertown. After checking the tank with a Geiger counter, Max (Mel Gibson) discovers it is radioactive to which the salesman replies "what's a little fallout?". This brief scene is a reference to the real-world dairymen in the 19th century who thinned milk with dirty water and embalming fluid, eventually leading to new safety regulations.[4] Later in the movie, Max explains how precious the clean water is to the "lost tribe" who are living under a waterfall. Water then becomes the dominant theme in *Mad Max: Fury Road* (2015), where a dictator is hoarding an aquifer of fresh water.

Sticking with the theme of water, next we have *Waterworld* (1995). The protagonist (Kevin Costner) uses a mechanical device with lots of tubing to convert urine into fresh water. He can only use urine with the device because salt water is "too hard on the filter". Like the earlier *Mad Max* instalments, oil and petrol are also prominent here. The antagonist "smokers" live on an old oil tanker, refining the oil when they need and they "hate sails".

They also use a rusty car with a 'Nuke the Whales' bumper sticker to drive around the oil tanker.[†] The "smokers" are portrayed in contrast to wind turbines and batteries used on the protagonist's boat as well as on the atoll. There is also a demonstration of what appears to be electrolysis, which is used to make hydrogen and oxygen for the balloon they eventually use to find dry land.

Finally, we stick with water and electrolysis for the US sci-fi thriller *Chain Reaction* (1996), about the government coverup of a new clean energy source. The research takes place at the University of Chicago, where a machinist (Keanu Reeves) and a physicist (Rachel Weis) are working on an experimental hydrogen-based reactor. The use of this location is a reference to the first sustained nuclear reactor (Chicago Pile-1), by Enrico Fermi and his team at the same university as shown in *Oppenheimer* (2023). Later in the movie, a glassware setup with tubing is briefly shown at the secret government facility, which was filmed in the real-world Argonne National Lab located just outside Chicago.

Water is described as "two parts hydrogen, one part oxygen", and the lab consists of several computers around a metallic and glass tank with lots of tubes. A blue laser is directed at the centre of a bubble which causes "hydrogen separation" and the temperature is said to be "2 million kelvin". In reality, water splitting is easily achieved using a relatively small input of voltage. However, fusion and sonoluminescence are also mentioned during these scenes. Sonoluminescence is a real concept involving the emission of light from imploding bubbles when excited by sound. Single-bubble sonoluminescence was first reported in the early 1990s, which probably inspired the plot here.[5] In the movie, the researchers are using a laser to induce nuclear fusion through sonoluminescence. Real-world sonoluminescence research has only achieved around 20 000 kelvin to date but a million kelvin is theorised.[6]

[†] The phrase 'Nuke the Whales' is a satirical example of extreme environmental destruction. It originated in the late 1970s, with several punk bands using it in their lyrics (Supreme Pontiff and The Fleshapoids). However, it is most likely a humorous combination of the 'No Nukes' and 'Save the Whales' movements, the latter of which inspired the plot of *Star Trek IV: The Voyage Home* (1986). The phrase 'Nuke the Whales' also appears at the end of a song about saving seals in the comedy *Nice Dreams* (1981) and was later popularised by *The Simpsons* in the episode *Lisa's Date with Destiny* (1996).

The initial test in the movie is unsuccessful, giving us an example of 'learning from failure' when the lead scientist exclaims "we learned something very important today – we found another way that didn't work. We need to keep our minds open and try again". The next test is successful, using a specific frequency to stabilise the reactor. They state that the "spectrometer shows stable hydrogen production" and that they are getting "more power than we are putting in". However, if fusion was occurring, the gas produced would be helium, not hydrogen. It is therefore unclear if the reactor represents electrolysis or fusion. The success of the reactor leads to friction between the scientists and the investors about whether the discovery should be shared freely or controlled. As the scientists are attempting to share the discovery online, they are killed and the lab is destroyed through sabotage of the reactor to cover up the discovery.

The rest of the plot revolves around a conspiracy by a secret government organisation to steal and cover up the discovery, in addition to framing the protagonists for the explosion and murder. There is also a reference to one of the researchers previously "blowing up a science building", which makes him the prime suspect. Noteworthy also is a debate about the cost *vs.* benefit of funding scientific research, stating that the "incident in Chicago is tragic to be sure; sometimes that is the price we must pay to ensure our competitive edge in the future". This represents a throwback to several previous examples of this debate.

Finally, two years later we have *The Saint* (1997), a big screen adaptation of the 1960s TV series discussed earlier. The movie follows a thief (Val Kilmer) who is hired by a Russian oil oligarch to steal a revolutionary new cold fusion formula. Our electrochemist is introduced with a reference to the controversial experimental results published by Pons and Fleischmann in the real world, who claimed to have discovered cold fusion in 1989.[7] The scene takes place in the Oxford University chemistry teaching lab, which is filled with students in lab coats. The benches contain conical flasks with colourful liquids, but there are also sensible glassware setups scattered around the room with tripods and Bunsen burners.

We are then introduced to the female electrochemist (Elizabeth Shue) who is researching "the formula" for cold fusion. She holds up a large flask similar to the actual apparatus used by Pons and Fleischmann in their experiment and explains "when positively charged deuterons are attracted to the palladium cathode, they cram together... inside the cathode, until they fuse and create energy in the form of helium". In the real world, a deuteron is a nucleus of deuterium (heavy hydrogen) consisting of one proton and one neutron. She continues the lesson by making the case for cold fusion, claiming it will end pollution and that there is more energy in "one cubic mile of seawater than in all the known oil reserves on earth".

Later, the thief breaks into her apartment and discovers her notebook which contains the chemical formulae for molybdenum complexes, as well as the "molecular D_2 ground state wavefunction". In the background, there is some experimental apparatus set up in her apartment. Since this movie is set in the late 1990s, it would be hugely unprofessional to run experiments in your home. Therefore, it is most likely that the movie makers copied this concept from earlier movies like *The Love Test* (1935) and *The Man in the White Suit* (1951), without realising that home labs were no longer acceptable. On her computer he also discovers some accurate "electrochemical reactions in the fusion cell" (Figure 9.1), which are actually copied from the original paper by Fleischmann and Pons.[7]

$$D_2O + e^- \rightarrow D_{(ads)} + OD^-$$

$$D_{(ads)} + D_2O + e^- \rightarrow D_2 + OD^-$$

$D_{(ads)}$ represents deuterons adsorbed onto the palladium electrode surface.

$$D_{(ads)} \rightarrow D_{(lattice)}$$

Surface-adsorbed deuterons move into the palladium lattice, eventually overloading

the lattice and resulting in multiple site occupancy and fusion.

$$D_{(lattice)} + D_{(lattice)} \rightarrow {}^3He + neutron$$

Figure 9.1   The electrochemical and nuclear reactions in the fusion cell at the cathode, as seen in *The Saint* (1997). © Rysher Entertainment and Mace Neufield Productions.

The thief then pursues the electrochemist romantically, with the aim of obtaining the formula. He eventually seduces her and obtains the formula when she isn't looking. We then discover that the formula for solving cold fusion is the solution to the real-world 'Particle in a Box' problem. This is a common application of the quantum mechanical model to a simplified system consisting of a free particle with impenetrable barriers. There are also some brief scenes showing the lab belonging to the oil oligarch, with lots of empty glassware and tubing. Also worth noting, the chemist in this lab is described as "eccentric" and has a strong Christian faith.

9.5 QUADRUPLE BONDS

We last left Roger Moore as Bond doing his best impression of *Star Wars*. Now we enter the 1980s with *For Your Eyes Only* (1981), but it doesn't really contain any onscreen chemistry. Similarly, we only have an implied reference in *Octopussy* (1983), in relation to the blue-ringed octopus. This species uses a venom containing tetrodotoxin to kill, causing paralysis in its victims as previously mentioned in respect to *The Serpent and the Rainbow* (1988).

We then return to Sean Connery as Bond in *Never Say Never Again* (1983), to remake the story of *Thunderball* (1965) about the theft of nuclear weapons by Spectre. Interestingly, there is no gas used in this version; instead, the impersonator uses a fake contact lens to secure the weapons. Later, when Bond fails a training exercise, we get an explanation of 'free radicals', which are described as "toxins which destroy the body and the brain, caused by eating too much red meat and white bread, and too many dry martinis". In response, Bond offers to cut out the "white bread". In chemistry terms, free radicals are molecular species which can exist independently with an unpaired electron.

Later, during a fight scene at the health clinic, Bond and the assassin stumble into a lab. Conical flasks, beakers, and round-bottomed flasks filled with colourful liquids are shown along with a random bottle labelled "poison". The assassin picks up Bond and throws him through the shelfing unit of glassware. Bond retaliates by chucking a random breaker of liquid at the hitman's face which causes him a lot of pain. The assassin then falls backwards, impaling himself on the glassware with a measuring cylinder sticking out of his back. Bond checks the label on the beaker, which reads "Urine Specimen – James Bond".

We then return to Roger Moore again in *A View to a Kill* (1985) with a plot revolving around unspecified microchips, drugging racehorses, and some exaggerated geology. After a suspicious horserace, Bond discovers that Zorin (Christopher Walken) is implanting a microchip (a type of microfluidic device) in horses which releases steroids when remotely activated. Because the drugs are stored on the microchip until activated, it is not detected during routine drug tests. Zorin's lab is hidden underneath the stables and features several cabinets filled with chemicals. Bond breaks into the fridge to investigate, which contains sample vials, test-tubes, and other containers. Later, when Zorin outlines his plan to flood Silicon Valley by placing a bomb in the "key geological lock", he states "for centuries, alchemists tried to make gold from base metals. Today, we make microchips from silicon, which is common sand, but far better than gold".

Next, Timothy Dalton takes the reins in *The Living Daylights* (1987) in which he is sedated by chloral hydrate. This is the second time that the real-world sedative appears in a Bond movie, since it was previously mentioned in *From Russia with Love* (1963). Bond is correct when he says it tastes like mouthwash, but its effectiveness is much slower in the real world. We then return to the theme of recreational drugs in *License to Kill* (1989), in which Bond goes rogue on a revenge mission after his friend Felix and his wife are attacked by a drug lord. The cocaine is sold through an evangelist (Professor Joe) and his meditation retreat turns out to be a cover for the drug-processing facility. It is explained that their "product dissolves completely in ordinary gasoline, making it absolutely undetectable". This is a reference to the real-world method for extracting the drug from coca leaves which is done using petrol before cleaning and crystallisation. Although it would be detectable by several lab instruments, there could be issues related to the commonly used cobalt thiocyanate field test due to the dilution factor.

The lab is then shown filled with beakers, conical flasks, and test-tubes containing colourful liquids, but blue liquids are the most prominent. There are large containers, conical flasks joined with tubing as well as condensers, all filled with blue liquids. This is a reference to the real-world test for cocaine used by law enforcement and forensic labs called the cobalt thiocyanate test. As well as the cobalt compound, the test also uses glycerine, hydrochloric acid, and chloroform, and turns blue in the presence of cocaine and/or crack. Finally, the "chief chemist" demonstrates a precipitate forming in a beaker when "an 18% mixture of cocaine and ordinary gasoline is mixed with ammonium hydroxide". He then filters off the cocaine away from the gasoline using a mesh, which is another reference to real-world production. Overall, this is one of the most chemically accurate Bond movies and a prime example of the duality of chemists.

After a brief absence, we change Bond once again, with Pierce Brosnan taking over in *Goldeneye* (1995) and *Tomorrow Never Dies* (1997), neither of which contain onscreen chemistry. So, we move onto *The World is Not Enough* (1999) and return to the theme of energy with a story centred around an oil pipeline. The explosive start to the movie is claimed to be caused by the combination of urea with cellulose from bank notes, activated by a magnesium-strip detonator. They take some poetic licence with this by exaggerating several real chemistry concepts, but there's no indication as to what "chemical reaction" takes place on Bond's hand that tips him off. Interestingly, following the trend of female physicists onscreen during the 1990s, this movie features one who disarms nuclear bombs (Denise Richards).

After a brief reference to "hydrogen bombs made with tritium", later weapons-grade plutonium is depicted as two matt black hemispheres which are "safe to handle". This is somewhat true since 'weapons grade' mostly consists of Pu^{240}, an alpha emitter which doesn't penetrate skin. However, it will also contain impurities of highly radioactive Pu^{239}. The plot culminates on a nuclear submarine where the villain attempts to insert the plutonium rod into the reactor after removing the control rods. However, this wouldn't have any effect since weapons-grade plutonium is less reactive than reactor-grade plutonium, hence why it is safe to handle. At the end, the physicist claims "the hydrogen gas is too high", which ruptures the hull of the submarine.

Pierce Brosnan finishes his term as Bond in *Die Another Day* (2002), but, again, there isn't really any onscreen chemistry to speak of. We will come back to Bond later with Daniel Craig, for a return to form for onscreen chemistry references.

9.6 SCI-FI AND SUPERNATURAL

This era also gave us some re-interpretations of Mary Shelley's famous story. *Frankenstein Unbound* (1990) is about a time-travelling scientist from the future, who enlists the help of Frankenstein's monster to help him reverse the time-slip using the lighting from a storm to power "the laser". There is a brief scene shown in a lab with bubbling and colourful liquids in an assortment of glassware and tubes. Next, *Re-Animator* (1985) and *Bride of Re-animator* (1990) are loosely based on the H. P. Lovecraft stories. In these movies the scientist brings a specific person back to life, like that seen previously in *Maniac* (1934). Both movies feature medical labs with some glassware, but no real chemistry to speak of.

Next, we have another depiction of an old friend in *Jekyll and Hyde* (1990), this time with Michael Caine playing the titular role. Interestingly, this version is much closer to the original story, meaning there aren't any scenes of him researching the formula. By the time we see the lab filled with glassware, he has already discovered the formula and the antidote. There is one scene showing him mix a white "salt" with three colourful liquids from separation funnels. The antidote is labelled "reflux" which is correctly explained as "flowing backwards". The police also bring the "reflux bottle" to a technician who is unable to figure out what it is from his lab setup, saying "if the powder is acidic, it should turn this indicator red". Also, the debate here is less about 'good and evil' and more about whether 'chemicals' can alter a person, referencing psychology. Jekyll's lab notebook is also entitled *The Chemical Change in the Human Body*.

We then return to superheroes in *Dark Man* (1990), which pays homage to some movies mentioned earlier. The real-world advancements in skin grafts at the time inspire the story of a scientist (Liam Neeson) who is researching synthetic "liquid skin" using a type of 3D printer. The resilience and perseverance of scientists is depicted here again when their initial experiments fail. The conversations are mainly biological in nature, referencing glucose and DNA, while showing some images of cells under a microscope. The lab features common scientific equipment such as test-tubes, gas cylinders, fume hoods, and beakers *etc.*

The protagonist is attacked in his lab and disfigured by a mobster as they search for a document left there by mistake. The doctors treat his burn injuries by severing the nerves that allow him to feel pain, which has the side effect of giving him extra strength since he can't feel anything. However, in a throw-back to the *Invisible Man* (1933), this also makes him mentally unstable and borderline psychotic. He sets up a new lab in an abandoned sulfuric acid plant, resulting in a lab montage as he pours various liquids between containers and beakers. Also, in a throwback to *Edison, The Man* (1940), he repeats his experiments dozens of times during the montage as he attempts to perfect the synthetic skin. To enact his revenge on the mobsters, he then reproduces their faces to impersonate them.

The face-swapping continues in the US thriller *Suture* (1993), where we return to a black and white depiction of a lab. Here the new face is achieved through "advanced reconstructive surgery". There is a scene in a large lab as the police examine the fragments from a car bomb. As the camera pans around we can see empty volumetric flasks, test-tubes, and storage containers. In the background there are cabinets filled with glassware as well as fume hoods. Overall, it's one of the most realistic depictions of a lab on screen, but there is no mention of chemistry, or any lab skills shown. We finish the face-swapping theme with a brief mention of *Face-Off* (1997), where a detective (John Travolta) goes undercover as a criminal (Nicholas Cage) to infiltrate his organisation. But there is no real chemistry worth speaking off, with only some images of glassware and syringes shown.

Next, we return to supernatural chemistry in *Death Becomes Her* (1992), about an actress (Meryl Streep) who is obsessed with staying young. The elixir of life is a luminescent potion which grants eternal youth, but the body continues to decay. It is described as "a touch of magic in this world obsessed with science". This results in her plastic surgeon and mortician husband (Bruce Willis) taking on a type of Frankenstein role to repair her body with formaldehyde and spray paint in a scene with chemical containers.

We then return to some examples of space chemistry starting with *Gattaca* (1997), about a future society driven by eugenics. The movie's title is based on the first letters of the four nucleobases of DNA. The protagonist (Ethan Hawke) is born outside of the eugenics programme and faces genetic discrimination. To complete his dream of going into space, he meticulously scrubs away his genetic material daily. He also uses a home lab to prepare his fake fingertips and urine samples. Beakers, syringes, and sample vials are shown on the first known onscreen example of an 'glow top' bench, a concept used again later in *Star Trek: Beyond* (2016) and others. There are also examples of medical labs throughout.

In the same year, we also have *The Fifth Element* (1997), which doesn't contain any onscreen chemistry. It is included in this discussion through the use of the word 'element', which refers to the classical Greek elements of earth, wind, fire, water and aether or void (space or vacuum). However, these are probably better compared to what we now call 'states of matter', *i.e.* solid (earth), gas (wind), liquid (water) and plasma (fire). There are various definitions of the word 'element' and it has changed over time. However, the classical use is often confused with the modern use of the word in relation to the periodic table of elements, which is especially true for younger audiences. This will be discussed again later in relation to *Elements* (2023).

Also worth a brief mention here is *Contact* (1997). In a brief scene, the protagonist (Jodie Foster) mentions "hydrogen times pi" in reference to a signal received from space. This references the naturally occurring 'hydrogen line' or '21 cm spectral line' in radio astronomy, so multiplying that by pi would prove that the signal they are receiving is not natural. Also, before boarding the spacecraft she is given suicide pills in case she cannot return. The story for *Contact* (1997) was written by the famed physicist Carl Sagan and, like *Chain Reaction* (1996), includes a female physicist in the lead role. As mentioned previously, it is likely that some of the visual imagery and story was influenced by *This Island Earth* (1955).

Finally, we finish this section with some brief mentions. In *Men in Black* (1997), we briefly see some lab glassware which is destroyed by a super-bouncy ball set free by Jay (Will Smith). In the remake of *Godzilla* (1998), radiation is once again referred to as the cause of the mutation which created the Japanese monster. This version also opens with some real footage of nuclear testing in the Pacific Ocean. In *The Green Mile* (1999), the prison guards crush up some non-descript tablets to knock Wild Bill (Sam Rockwell) out so they can sneak John Coffey (Michael Duncan) out of the prison.

9.7 CORPORATE CHEMISTRY

We return to corporate chemistry with a rare example of a pharmaceutical company depicted onscreen as doing good for humanity in *Medicine Man* (1992). The plot follows a biochemist trying to isolate and extract "a cure for cancer" in the Amazon rainforest. The pharmaceutical company funding the research sends another researcher (Lorraine Bracco) to his lab in the jungle with a gas chromatograph (GC) at his request. In addition to some conical flasks and reflux condensers, some gas cylinders are noticeable also. However, it is never explained what is powering any of the systems in such a remote location. Campbell (Sean Connery) runs his plant extract through the instrument, but Crane (Lorraine Bracco) correctly asks him if he has "run a baseline to calibrate the machine" and tells him that there is a glucose solution for that (labelled with the correct chemical formula).

However, a detector is not mentioned, but the spectrum provided on the computer gives them a list of 49 compounds. The computer automatically gives her the chemical formulae for all the compounds, which does not happen in real life. On the list are some real substances like Na_3PO_4, $Mg(HCO_3)_2$ and KCl, but the majority are not volatile enough to show up on a GC. It also lists a non-existent substance like $NaCl_2$. As Crane beings to rain on his parade, the rain in the jungle also picks up pace. She then clicks on the "unidentifiable" peak number 37, which gives us a large multi-ringed complex which "cannot be synthesised" because it is described as "mother nature's kitchen" (Figure 9.2).

Later, we discover that a profit-driven logging company is building a new road to the village, exposing the native population to outside pathogens. Campbell pleads to halt the construction until he can conclude his research. After a string of

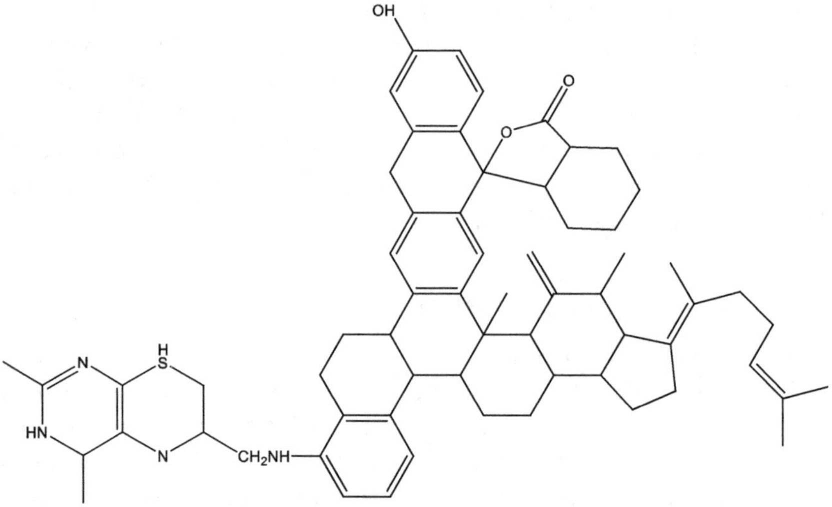

Figure 9.2 The cure for cancer as seen in *Medicine Man* (1992), discovered using a gas chromatograph in the Amazon rainforest. © Hollywood Pictures and Cinergi Pictures.

unsuccessful attempts, they accidentally discover that the source of the cure is not a flower but a species of rare ant indigenous to the rainforest. In their desperation, the chemists end up spreading a fire which destroys the research station and the surrounding rainforest. However, the pharmaceutical company increases their resources to continue their work and replaces the lab.

Next, we have two of the most extreme examples of unscrupulous pharmaceutical companies. First, we have *The Fugitive* (1993) in which the protagonist (Harrison Ford) is a medical doctor who is framed for his wife's murder. It later transpires that a pharmaceutical company is behind the murder because the protagonist discovered that their new drug (Provasic) causes liver damage. One of the board members of the company covered up the side effects, by replacing unhealthy samples with healthy ones, to get it approved by the FDA. In a similar vein we also have *I Love Trouble* (1994), about a chemical company who causes a train crash to hide evidence and kill the people who were about to expose the cancerous side effect of their new livestock milk-production drug. This fictional movie hints at the real-world controversy around the use of bovine somatotropin (BST) in US milk production. Also worth a brief mention here is the comedy *Kids in the Hall: Brain Candy* (1996), which mocks corporate chemistry culture and marketing through the rapid and careless development of a new anti-depressant.

Next, we have *The Insider* (1999), which is a fictionalised account of a true story about a whistleblower biochemist (Russell Crowe) from the tobacco industry. He is bound by a confidentiality agreement which he initially abides by, but then receives death threats. Eventually, he agrees to do an interview where he claims that the tobacco companies were using "ammonia chemistry, which allows for the nicotine to be more rapidly absorbed in the lungs". This is true, since ammonia can alter the natural acidity of nicotine to help this process. He goes on to talk to about the real flavour enhancer coumarin and its use in tobacco, which he describes as a "lung-specific carcinogen". It was banned as a food additive by the FDA in 1954.

Of note, he mentions several real pharmaceutical and health-related companies as his previous experience. There is also a scene in a chemistry teaching lab, where he now works after been fired from the tobacco company. He tells the students that he "finds chemistry to be magical, I find it an adventure. An exploration into the building blocks of our physical universe". He begins the lesson by telling them that their first experiment will be "measuring the molecular weight of butane", which is a real student experiment.[8] Later, he also writes the correct equation on the chalkboard for the decomposition of potassium chlorate. These scenes represent one of several examples of teaching labs in movies and on TV in the late 1990s, a theme which will be discussed later.

Finally, we finish here with the drama thriller *The Constant Gardener* (2005), about a British diplomat (Ralph Fiennes) living in Kenya who attempts to solve the murder of his wife (Rachel Weisz), an Amnesty International activist. During his investigation it transpires that she found evidence linking deaths to trials of a new (fictional) anti-tuberculosis drug called Dypraxa. This leads to a coverup which connects her death to a damming report she was preparing about the drug. The drug trials were carried out by a subsidiary of a large pharmaceutical conglomerate, and it is implied that they are responsible for death threats, torture, and murder. It later transpires that the report was silenced to save the company millions in redeveloping the drug. Eventually, the implied becomes explicit when henchmen belonging to the company murder the protagonist and make it look like suicide at the same spot as his wife. The coverup is finally exposed at their memorial.

The plot of the movie is inspired by real-world drug trials previously carried out by pharmaceutical companies in several African nations. The movie also references public complacency regarding the human cost of modern medicine. However, there is of course a logic flaw in the plot, in that, if a drug is causing so many deaths, how does a company plan to market such a drug and continue to cover up its lethality? The argument made onscreen is that it was too costly to develop, and it can't be sent back to the lab. This doesn't make sense either, since it would cost more to continue the trials with a flawed drug. As a result, *The Constant Gardener* (2005) represents a missed opportunity to properly explore lax standards when companies do trials on the cheap.

REFERENCES

1. C. J. Cambell, *Emperor Nero's Death & The Curious Case Of The Pseudo Neros*, The Collector, https://www.thecollector.com/emperor-nero-death-pseudo-neros/.
2. J. Hare, *Acid lakes: do they exist and would they dissolve a boat?* Education in Chemistry, https://edu.rsc.org/opinion/acid-lakes-do-they-exist-and-would-they-dissolve-a-boat/2021311.article.
3. J. Hare, Soap: Can you make it with body fat and is there an explosive spin-off?, Education in Chemistry, https://edu.rsc.org/opinion/soap-can-you-make-it-with-body-fat-and-is-there-an-explosive-spin-off/2021319.article.
4. Deborah Blum, The 19th-Century Fight Against Bacteria-Ridden Milk Preserved With Embalming Fluid, Smithsonian Magazine, https://www.smithsonianmag.com/science-nature/19th-century-fight-bacteria-ridden-milk-embalming-fluid-180970473/.
5. L. A. Crum and S. Cordry, Single-Bubble Sonoluminescence, in *Bubble Dynamics and Interface Phenomena*, ed. J. R. Blake, J. M. Boulton-Stone and N. H. Thomas, Springer Netherlands, Dordrecht, 1994, pp. 287–297.

6. W. C. Moss, D. B. Clarke, J. W. White and D. A. Young, Hydrodynamic Simulations of Bubble Collapse and Picosecond Sonoluminescence, *Phys. Fluids*, 1994, **6**, 2979–2985.
7. M. Fleischmann and S. Pons, Electrochemically Induced Nuclear Fusion of Deuterium, *J. Electroanal. Chem.*, 1989, **261**(2 PART 1), 301–308.
8. Nuffield Foundation, Determining the relative molecular mass of butane, Education in Chemistry, https://edu.rsc.org/experiments/determining-the-relative-molecular-mass-of-butane/1720.article.

Plot Devices

10.1 DANGEROUS CHEMICALS

At the beginning of the 1990s, new safety measures and protocols reduced the frequency and seriousness of industrial accidents. However, the theme of greedy and evil corporations
continued to feature onscreen, with real-world accidents inspiring fictional versions. This resulted in the word 'chemicals'
entering the common lexicon for anything dangerous and
harmful to one's health, often serving as a plot device. For instance, the word appears very briefly in *Wrong Turn* (2003) when
"spilled chemicals" are mentioned as the reason for a traffic jam,
forcing the protagonists to find an alternative route with deadly
consequences.

Similarly, we also have *Sahara* (2005), about pollution in the
Niger river causing disease. Even though the terms "toxic waste",
"industrial waste", "chemical waste" and "biowaste" are inter-
changed randomly throughout the movie, it is never explained
what the substance is. However, we can assume it is biological in
nature since it is causing a "plague" and they claim the growth
rate will multiply when it "hits the salt water in the ocean".
Eventually the protagonists discover that the corporate antagonist has built a solar processing plant in the desert to va-
porise the waste. But it isn't fully operational, so he is storing the

Onscreen Chemistry: The Portrayal of Chemical Science in Film and TV
By John O'Donoghue
© John O'Donoghue 2025
Published by the Royal Society of Chemistry, www.rsc.org

barrels underground in a cave. The barrels are shown labelled with biohazard symbols.

Worth a brief mention here also is a remake of *Total Recall* (1990), which itself is a screen adaption of the 1966 story *We Can Remember It For You Wholesale* by Phillip K. Dick. *Total Recall* (2012) is set at the end of the 21st century, when chemical warfare has devastated the Earth, leaving only two habitable lands on either side of the planet. However, details about the chemical warfare are vague and unspecific.

10.2 CARTOON ENVIRONMENTS

The environmental cost of persistent chemicals wasn't just confined to live action depictions, it also started to make its way into animated onscreen chemistry as well. Following our previous discussion about *Scooby Doo* (1969–present) and *The Muppets* (1955–present), we also have several cartoon depictions of chemistry from the 1980s and 1990s with similar themes. We'll start here with a brief mention of the *Teenage Mutant Ninja Turtles* (1987–present), which made the jump from comic books to the screen a year after the 1986 Chernobyl nuclear accident.

The glowing green "ooze" mutagen which creates the turtles and their master Splinter is sometimes said to be from another dimension. However, it is most often associated with the improper disposal of chemical waste. In the fourth live action big screen adaption of their origin story, *Teenage Mutant Ninja Turtles* (2014), the mutagen originates from a lab. The concepts of mutations and chemical waste continue with the US environmentalist superhero series *Captain Planet and the Planeteers* (1990–1992). Here, the team are often found cleaning up chemical waste and stopping polluters which they call "eco-villains". The series also includes another reference to *Dr Jekyll and Mr Hyde* in relation to one of the recurring villains. Those familiar with video games might find it interesting that the villain is a doctor who has changed himself into a radioactive mutant called Duke Nukem.

The Captain Planet series usually revolves around teamwork, with their combined powers solving many of the issues that they are presented with. A wide range of themes are explored, including overfishing, over-population, designer viruses, urban development and more. Many of the chemistry-related themes echo discussions from previous seasons, such as drilling for oil, acid rain, mining, water contamination, nuclear waste, global warming, and clean energy. However, there are also several depictions of scientists in negative roles too, such as Duke Nukem attempting to destroy the ozone layer in *Ozone Hole* (1991). Another example involves scientists covering up a radiation leak in a nuclear power plant in *Meltdown Syndrome* (1991). In a throwback to *The Man with the Golden Gun* (1974), there is also an episode featuring a scientist who has perfected devices for storing and transmitting solar power, but Duke Nukem attempts to subvert this.

However, relevant to this discussion about onscreen chemistry, as well as dealing with environmental issues, the series also tackled the issues of drug abuse in the episode *Mind Pollution* (1991). The drug here is a fictional designer drug called 'bliss' and causes one of the characters to jump through a window and die of an overdose. The series also dealt with the HIV/AIDS pandemic in the episode *A Formula for Hate* (1992), which revolves around brainwashing.

Sticking with 1990s animations, next we have *The Simpsons* (1989–present), about a suburban family who live in the US town of Springfield. The father, Homer, works in the local nuclear power plant which uses a pressurised water reactor, as explained in *Homer Defined* (1991). However, the nuclear plant is depicted as having very lax safety measures. Glowing green fuel rods are usually found lying around the canteen and Homer's frequent carelessness is mocked by Frank Grimes in *Homer's Enemy* (1997). Ironically, Homer obtains his job as safety inspector when he becomes an anti-nuclear safety activist after he was originally fired for gross incompetence.

Sticking with the environmental theme, the green liquid waste from the Springfield nuclear plant is often shown in various locations around the town and causes mutations like the three-eyed fish seen in *Two Cars in Every Garage and Three Eyes on Every Fish* (1990). Later, the episode *King-Size Homer* (1995) features an inexplicable toxic gas, which Homer stops by blocking the hatch because he is overweight. In real life, nuclear waste is not liquid, spent fuel rods remain as a ceramic solid, but they are stored in water for many years. This is possibly where the misconception arises.

Over the course of the long-running series there are also several other science references such as a couch gag which zooms out to the entire universe before returning to atoms, molecules, DNA and back to the couch. There are also several 'film-within-a-film' educational videos, usually shown *via* projector in the school and presented by the character Troy McClure. Some include onscreen chemistry like *A World Without Zinc*, which provides real-world examples of uses for the chemical element zinc like car batteries, cars, and guns.

There is also a school education video about DNA, which includes laboratory glassware. Troy pours some pink liquids into a beaker to simulate the genetic contribution of both parents and finishes by tasting it. However, the letters 'DNA' are never explained as **D**eoxyribo**N**ucleic **A**cid. There is also a film within a film shown to the school students when they visit the nuclear plant on a field trip. The film is entitled *Nuclear Energy – Our Misunderstood Friend* in which Smiling Joe Fission talks positively about nuclear energy with "rods of uranium-235". However, near the end of the video, he also sweeps "leftover nuclear waste" under a rug. There is also a brief video about sand which correctly explains that it can be used to make windshields. Finally, there is also one entitled *Birds* which features a lab containing glassware filled with colourful liquids.

Importantly for the discussion here, the Simpsons franchise also contains a recurring 'mad scientist' in the form of Prof. Frink (Figure 10.1), who is based on the Jerry Lewis version of *The Nutty Professor* (1963). In addition to his typical nerdy eccentricities, he is nearly always wearing a lab coat, as well as a bow tie and thick round glasses. Like Bunsen Honeydew and Beaker in *The Muppets* (1955), he is normally depicted as working on bizarre inventions which usually do not work, are of no real use, or makes things worse. Interestingly, he was originally depicted as an 'evil scientist' in his first appearance in the episode *Old Money* (1991) when he tries to secure funding for a "death ray".

However, following this episode his intentions have nearly always been noble, since he is usually depicted as trying to help the town of Springfield. Some examples of his inventions include hamburger earmuffs, the gamble-tron 2000 which predicts that a football team will win by 200 points, and a burglar-proof house which sprouts legs and runs away. Noteworthy in terms of

Figure 10.1   Professor Frink from *The Simpsons* (1989–present) is modelled
after Jerry Lewis from the original *The Nutty Professor* (1963).
© Photo 12/Alamy Stock Photo.

onscreen chemistry, it is also claimed that he has discovered the
fictional element frinkonium. Also of note is the special episode
Treehouse of Horror XIV (2003) in which Prof. Frink revives his
dead father in a homage to Frankenstein. In the episode, his
father is voiced by Jerry Lewis as the character Julius Kelp from
The Nutty Professor (1963).

Interestingly, chemistry is also used to solve a problem in the
episode *Bart the Murderer* (1991), when the school principal,
Seymour Skinner, goes missing and is presumed dead. During
Bart's court case for the crime of his murder, a dishevelled
looking Skinner bursts into the room and explains what happened to him. It turns out that while cleaning out his garage he
got trapped under a pile of newspapers. To get out from under
the pile, he makes a crude rocket from a discarded cigar tube
and "remembering an experiment from my time as a 4th grade
science teacher, I concocted a fuel from baking soda and some
discarded lemon wedges". He then goes on to explain that the
"rocket took off with a mighty blast of carbon dioxide, dragging
with it the end of a vacuum cord", which is chemically correct.

However, it is very doubtful that the rocket would have enough thrust to pull the cord and form a support line for Skinner to save himself. Nonetheless it is commendable in its chemical accuracy.

We finish this section with the environmentally themed animated sci-fi romantic adventure *WALL-E* (2008). Set in the 29th century, the story follows the last functional trash-compacting robot named WALL-E (Waste Allocation Load Lifter: Earth-class) left on earth to tidy up the huge mountains of waste left behind by the humans who have fled. All the scenes on Earth have a brown tinge due to the build-up of nitrogen dioxide, a common pollutant from burning fossil fuel. Interestingly, WALL-E uses a solar panel to recharge batteries, so most of the environmental message is around consumerism due to the large quantities of waste.

10.3 CARTOON CHEMICAL PLOT DEVICES

We now return to the Acme Corporation, the fictional supplier of everything the Coyote needs to catch the Road Runner. Following the theme of 'evil industry' that gained prominence from the 1970s onwards, Acme also started to evolve into the fictional cartoon version of this theme. There are two satirical stories about lawsuits against Acme during this era. In 1982, the *National Lampoon* magazine featured a three-part series of articles called Cliff-Hanger Justice by Joey Green, which gave a fictional account of a product liability lawsuit by Wile E. Coyote against Acme. Later, Ian Frazier also wrote a fictional legal complaint *Coyote v. Acme*, which was published in *The New Yorker* in 1990.

Back onscreen, the plot of *Who Framed Roger Rabbit* (1988) revolves entirely around the Acme Corporation. Set in the late 1940s, 'toons' co-exist with humans in Los Angeles where a portal links the live-action real world to Toontown. In the movie, the fictional studio owner hires a private detective (Bob Hoskins) to investigate if Roger's wife Jessica is having an affair with Marvin Acme, owner of the Acme Corporation. Later, Roger is framed for Marvin's murder in the Acme factory/warehouse, which is filled with wacky products. The slogan for the Acme Corporation is also mentioned by Marvin and displayed on the outside of the factory as "If it's Acme, it's a gasser".

Toontown has a human-looking judge named Doom (Christopher Lloyd) who executes toons using a dip consisting of "turpentine, acetone, benzene" which can destroy the otherwise invincible toons. After the evidence points to Roger, Doom plans to use the dip on him as well. The metal drum containers are labelled with a yellow corrosive symbol and sometimes contain the word "turpentine" while other times they are labelled "methylated spirits". Later we discover that Doom actually killed Acme so he could destroy Toontown using a giant container of the dip. Finally, it transpires that Marvin's last will was written using "disappearing, reappearing ink" and states the toons inherit Toontown.

Following *Who Framed Roger Rabbit* (1988), the Acme Corporation transitions further towards the 'evil industry' theme when it appears again as the antagonistic force of *Looney Tunes: Back in Action* (2003). Here, the head offices of Acme are shown as part of a multinational faceless corporation whose executive officers are led by the main antagonist of the movie (Steve Martin). The plot revolves around an Acme-led plan to transform the world's population into subservient monkeys using the "Blue Monkey diamond".

Next, we head over to Acme Labs, home to *Pinky and the Brain* (1993–1998). The mouse duo first appeared as a recurring segment in the *Animaniacs* (1993–1998) TV series but were later spun off into their own show due to their popularity. The series follows two genetically engineered mice who break out of their cage every night and "try to take over the world". There is also an interesting reference to the duality of science in terms of the description of the two characters with one described as a "genius", while the other is called "insane". The Brain is highly intelligent but also self-centred, while Pinky is good natured but often depicted as bumbling and feeble. As mentioned previously in relation to *The Incredible Shrinking Woman* (1981), there are plenty of scenes showing the lab from their perspective, *i.e.* everything including the glassware is oversized.

The series also features other "genetically spliced" recurring characters, such as snowball the hamster and Billie the female white mouse who is more intelligent than Brain. In *Snowball* (1996), the tables are turned when the mice save the world from the evil hamster so they can take over the world the following night. The plot device is usually based around raising money to finance their plans, once again referring to the theme of funding scientific research. The concept of conspiracy theories also appears in *Pinky Protocol* (1997), when a Hollywood director claims that two laboratory mice are trying to take over the world.

From a chemistry perspective, in *Pinky and the Brainmaker/ Calvin Brain* (1997) they create a mind-controlling fragrance. The human lab scientists also appear from time to time, to perform tests on the mice. In one example, Brain is subjected to experiments to examine the addictiveness of cigarettes in *Inherit the Wheeze* (1998), referencing similar topics to that seen a year later in *The Insider* (1999). There are also several references to schools across various episodes. These include a homage to the movie *Dangerous Minds* (1995) in the episode *Dangerous Brains* (1998), when Brain takes a teaching job at an inner-city school to secure financing for world domination. This theme will be discussed further in the next season.

Following this we have a cartoon homage to *Charlie's Angels* (1976–1981) in *The Powerpuff Girls* (1998–2005), featuring Professor Utonium, creator of the superpowered girls. Every episode opens with a sequence in the lab explaining that professor Utonium was attempting to create the "perfect little girl" using a mixture of "sugar, spice, and everything nice". However, he accidentally spilled a substance referred to as "Chemical X" (also called "Chemical Xtreme") into the mixture, thus granting them superpowers which include flight, super-strength, super-speed, X-ray vision, red heat vision, energy projection, and thermal resistance, among others.

Many of the plots revolve around the girls recovering stolen Chemical X from their archnemesis Mojo Jojo, the professor's mutated pet monkey. Mojo Jojo attempts to use Chemical X to turn other primates evil and intelligent in his image, with oversized brains, green skin, pink eyes, and the ability to speak. Chemical X is described as a powerful mutagenic chemical that gives people and animals superpowers and special abilities. It is typically shown as a viscous black liquid like pitch oil and is usually contained in a round-bottomed flask sealed with a cork. However, in the episodes *Mo Job* (1999) and *Twisted Sister* (1999) it also appears as a glowing blue liquid, and in *Toast of the Town* (2003) it appears as a powder. The effects of Chemical X can also be counteracted using Antidote X, which is synthesised in a lab. It can also be removed from the girls which causes them to lose their powers. The use of fictional materials as plot devices will be discussed in more detail later.

Finally for this section, we have *Dexter's Laboratory* (1995–2003), about a boy-genius with a hidden science lab in his room. Dexter is always shown wearing a lab coat, round glasses, and gloves. He doesn't really have many hobbies other than his lab inventions, something that his sister Dee Dee points out to him in *Way of the Dee Dee* (1996). Many of the plots revolve around his competitiveness with his classmate who is also a boy-genius. Dexter is usually at the losing end because his older sister Dee Dee inevitably foils the experiments when she gains access to the lab. Though seemingly dim-witted, Dee Dee is a talented ballet dancer and often outsmarts her brother, as well as offering helpful insights from time to time. Like *The Powerpuff Girls* (1998–2005), there is also a lab monkey and frequent references to well-known superheroes. The lab is also shown with some lab glassware, but most of his inventions are technology based.

10.4   MAD SCIENTISTS

Previous discussions saw how scientists were called 'mad' or 'insane' due to their persistence and obsession, or because of their unusual ideas. The movie *Maniac* (1934) also provided us with one of the first explicit images of mad behaviour coupled with some of the imagery that we now associate with 'mad scientists'. Following the early depictions of scientists onscreen, the 'mad scientist' image slowly developed during the gothic horror and monster movies of the 1940s and 1950s. It almost became the full package for Jerry Lewis's *The Nutty Professor* (1963), albeit without the wild white 'Einstein' hair.

A decade later we get the full 'mad scientist' image in Gene Wilder's manic depiction of Victor Frankenstein's grandson in the comedy horror *Young Frankenstein* (1974), complete with wild hair and exaggerated round safety goggles. This image was then cemented by Christopher Lloyd's "crazy, wild-eyed scientist" Doc Brown in the *Back to the Future* (1985–1990) trilogy. Doc Brown is the complete package of what is generally accepted as the 'mad scientist' image,[1-4] with wild white hair, lab coat, occasional round safety goggles and large yellow gloves as he handles plutonium.

Previously we discussed Prof. Frink in *The Simpsons* (1989–present) franchise as a homage to Jerry Lewis's *The Nutty Professor* depiction. We also see evidence of this image in *Futurama* (1999–present), which follows the cryogenically preserved Philip J. Fry. The protagonist wakes up 1000 years in the future and joins a delivery company which is run by his 160-year-old nephew Professor Hubert J. Farnsworth. Commonly referred to as 'the professor', Farnsworth shares a lot of similarities with Prof. Frink from the Simpsons franchise. Like Prof. Frink, he spends much of his time inventing ridiculous devices and contraptions, which usually fail or result in unwanted side effects. However, instead of general clumsiness, much of Farnsworth's mishaps are usually associated with age-related senility.

In the episode *Mars University* (1999), when asked what he is teaching, he responds: "The same thing I teach every semester: the mathematics of quantum neutrino fields. I made up the title so no student would dare take it". He rarely worries about his delivery crew, only viewing them as a means to an end. There is also an explicit reference to the H. G. Wells novel *The Sleep Awakes* in the episode *A Fishful of Dollars* (1999), when Fry discovers his bank account has continued to accrue interest over the course of a thousand years.

More recently we also have a parody of the *Back to the Future* trilogy with the US adult animated sci-fi series *Rick and Morty* (2013–present). The 'mad scientist' here is Rick Sanchez, a misanthropic and alcoholic scientist known for his reckless behaviour. Rick's lab is in the garage and features a laboratory glassware setup in the background of nearly episode. In one episode he makes "ovenless brownies" by mixing "titanium nitrate and a tad of chlorified tartrate" between test-tubes. Titanium nitrate is a real colourless and volatile solid, but the other substance doesn't exist. Tartrate is the salt of cream of tartar.

Finally, worth a brief mention here also is the "unlicensed scientist" Dr Krieger in the comedy animated spy series *Archer* (2009–2023), who "never formally earned his doctorate". He is nearly always shown wearing a lab coat and is usually depicted as being mad, underhanded, sadistic, manipulating, but also good natured on some occasions. Like the other cartoon scientists here, he spends his time inventing strange devices or carrying out unethical experiments. The chemical references usually revolve around nerve gas and psychotropic drugs like his "Magic Breath Strips" which later turn out to be laced with LSD. Another example is the "Krieger Kleanse", a tea for evading drug tests which turns out to be a hallucinogenic experiment.

10.5   SUPERNATURAL SCIENCE

In this section we will now look at how the supernatural genre has evolved after the turn of the millennium, starting with a modernisation of the famous H. G. Wells story about invisibility. *Hollow Man* (2000) opens with the protagonist (Kevin Bacon) using computational chemistry to "stabilise" a 3D molecular structure on his computer. The basement government lab where he works contains animal test subjects which include a gorilla who is already invisible but very aggressive. The research team are trying to reverse the invisibility, which becomes the focus of the plot. Infrared cameras are used to see the specimens when they are invisible.

The word 'serum' is used here in contrast to the word 'formula' used in earlier onscreen chemistry depictions. The "protein serums" are created in a type of sealed chamber. The deputy lead researcher (Elizabeth Shue) places a vial of colourless liquid in and out pops an "irradiated" glowing liquid. The blue version is used to go invisible, while the red version brings them back, *i.e.* opposite that of the same-coloured pills in *The Matrix* (1999) released a year earlier. The invisibility is explained using some nonsensical terms like "phase shifting out of quantum sync with the visible universe" and coming back is called "a quantum reversion". Against the advice of his colleagues, the protagonist decides to proceed with human testing. He proclaims "you don't make history by following the rules – you make it by seizing the moment", which is a homage to a similar discussion in *Monkey Business* (1952).

When they return to the lab there is another interesting example of the link to madness when the other researchers call the protagonist "insane" for volunteering to be the first human trial. The deputy lead researcher states "Jonas Salk tested the polio vaccine on himself, was he insane?", to which a colleague responds "ya, I'm pretty sure he was". Later there is also a throwback to the obsession and persistence tropes when the term 'workaholic' is used. The invisibility process works on the protagonist but the aggression side effect hinted at earlier exacerbates the worst parts of his personality, which is a reference to *The Invisible Man* (1933). The movie finishes by returning to the 'chemistry explosions' trope when he uses sulfuric acid and some other unnamed liquids to create an explosive described as "nitro". This is most likely nitroglycerin, discussed earlier, which does require sulfuric acid for production.

Worth a brief mention is another modernisation of the story in *The Invisible Man* (2020), described as a reboot of the original franchise that finished in 1951. It opens with a pill bottle labelled "Diazepam", a real-world sleeping pill, which the protagonist uses to escape an abusive relationship with Griffin. Following her escape, Griffin fakes his death and turns invisible to enact his revenge. Later, a doctor tells her that she has diazepam in her system, and she finds the same bloodied diazepam pill bottle from earlier in her bathroom. The invisibility here is achieved using an "optical suit" that the person wears. The only other chemistry reference is Griffin's optics company, which is called "Cobolt", a misspelling of the chemical element cobalt.

Next, we return to food chemistry in the Tim Burton musical *Charlie and the Chocolate Factory* (2005). The movie title is the same as the original 1964 Roald Dahl novel, unlike the previous screen adaptation which emphasised Willie Wonka over the protagonist Charlie. Later, there is also an origin story about the chocolate maker called *Wonka* (2023), which contains some brief scenes with glassware. The 2005 movie opens with Charlie's father working in a toothpaste factory, with "menthol" labelled on the large metal tanks in the background. Menthol is a naturally occurring chemical found in peppermint and other mint plants. It was first added to tobacco in the 1920s to reduce the 'harshness' of cigarette smoke and nicotine.[5]

Against the odds, Charlie finds a golden ticket, giving us a look inside the mysterious factory. Unlike the 1970s version, there is no scene in a classroom lab, but there is a brief scene in the factory's "inventing room" showing laboratory glassware. All the glassware contains brightly coloured liquids, which includes conical flasks, test-tubes, boiling flasks, and funnels. Some are also bubbling and emitting smoke/condensate. This area is described as the "the most important room in the factory", re-linking chemistry with the theme of invention. In the same room, Wonka also picks up a piece of "hair toffee" from a colourful and elaborate chemistry glassware apparatus with lots of tubing.

A year later we return to monster mayhem with the South Korean supernatural 'anti-chemical waste' movie *The Host* (2006). The story begins when a US military pathologist orders his assistant to dump hundreds of bottles of formaldehyde down the drain because they were covered in dust. His assistant warns him that they are "toxic chemicals" and regulations state that they need to be disposed of properly, but the pathologist insists. As the assistant pours it down the drain, he wears a mask as a large amount of vapour is produced.

Formaldehyde is the older name for methanal and is used as a preservative for organic tissue. It is also a volatile organic compound (VOC), meaning it can easily become a gas at room temperature, which makes this scene accurate. It is also considered a mutagen so there is some basis in real science when the movie characters find mutations in fish life, eventually resulting in a large monster. Later, there are also various scenes with hazmat suits since the monster is claimed to be the host of a deadly virus. However, it is discovered that this was part of the coverup to hide its origins. The military eventually use mustard gas to stun the beast, also causing blisters on surrounding people which happens in real life as well.

We then move onto the mind-bending French-Canadian fantasy love story *Upside Down* (2012), about an alien world with "dual gravity". The two worlds can see each other with one literally above the other. As a teenager, the protagonist (Jim Sturgess) from the "Down Below" eats "flying pancakes", made using pollen collected by pink bees from both worlds. His love interest (Kirsten Dunst) is based in the "Up Top" and they meet each other at the top of a mountain on both worlds. Later, he perfects an anti-gravity cosmetic product for literal face-lifts, using lab glassware like conical flasks, funnels, condensers, and beakers, all filled with bright pink liquids. This eventually leads to a "formula" which allows people to live on both worlds, so his chemical discovery makes their relationship possible.

Next, we have a homage to *Re-Animator* (1985) with the medical horror *The Lazarus Effect* (2015). The plot follows medical researchers who develop a "serum" to bring people back from the dead, which also leads to a subplot about science *vs.* religion. The basement lab contains glassware and microscopes as they explain how the serum works in conjunction with electrical shocks to "jump start the brain". However, they discover that bringing people back from the dead has some unforeseen side effects. A pharmaceutical company then acquires and confiscates all their work. While trying to prove their ownership, one of the researchers is accidentally electrocuted, so her colleagues use the serum to bring her back. However, the serum causes her brain to rapidly "evolve" and she develops superhuman abilities like telekinesis and telepathy.

10.6 FICTIONAL MATERIALS

As discussed previously, fictional elements or minerals provide writers with a license for creativity and often appear as a convenient plot device to move a story forward. Previously we discussed the fictional element nitron in the *Flash Gordon* serials (1936–1940) and discovered the fictional element meteorium in the *Lost City of the Jungle* (1946), which had an "an atomic weight of 245". Around the same time, Duffy Duck is sent on a mission to find "illudium phosdex" in *Duck Dodgers in the 24$\frac{1}{2}$th Century* (1953). He is told that the world's supply of the "shaving cream atom" is low. After following the planets in alphabetical order, he eventually discovers a source of the material on "Planet X". We also discussed the elemental analysis of the fictional mineral kryptonite in *Superman III* (1983) and mined for turbidium on Mars in *Total Recall* (1990).

We will now discuss other fictional minerals and elements that are usually central to a plot. Another well-known example of an important fictional material is the "dilithium crystals" from the *Star Trek* franchise. They first appear onscreen in *The Alternative Factor* (1967) from the original *Star Trek* (1966–1969) series. However, their first in-universe appearance is in the episode *Cold Front* (2001) from the *Star Trek: Enterprise* (2001–2005) series when Tucker is asked "How do you regulate positron flow in your dilithium matrix?". Note, since this episode is set very early in the *Star Trek* universe, the engines are not called "warp drives", they are called a "gravimetric field displacement manifold".

It is claimed that dilithium serves as a controlling agent for matter–antimatter reactors like those used for warp drives. It is also explained that warp drives use a stream of deuterium gas (heavy hydrogen) as the source of matter and anti-deuterium as the source of anti-matter. The annihilation fusion reaction heats the excess deuterium gas producing "warp plasma", which facilitates faster than light travel. In the original series, dilithium crystals were rare and could not be replicated, making the search for them a recurring plot device.

In the episode *Rascals* (1992) from *Star Trek: The Next Generation* (1987–1994), a fictional extended periodic table of "hypersonic elements" is briefly shown with dilithium appearing under francium, with the atomic number 119 and chemical symbol Dt. The atomic number 119 would correctly place it at the bottom of the first group (alkali metals), under the real element francium. Since lithium is the first element of the alkali metals, this is also the most likely source of the name.

In the rebooted (*Kelvin* timeline) *Star Trek* movies, which include *Star Trek* (2009), *Star Trek into Darkness* (2013) and *Star Trek Beyond* (2016), the large tanks in the warp drive contain an unusual 1940s version of the radiation symbol (trefoil) with outwards arrows placed between the blades. Since francium is radioactive, it is likely that dilithium is also radioactive so perhaps this is the reason for the ionising radiation symbol. Also, worth a brief recall here is the underlit lab bench containing glassware with colourful liquids in the *Star Trek Beyond* (2016). The concept of a lab bench with an under-light was last seen in *Gattaca* (1997), and remains an unusual visual concept for onscreen chemistry.

Sticking with the topic of the periodic table, we also see an extended version in the background of the school classroom in the *Snowpiercer* (2020–2022) series. This series is based on the 2013 movie of the same name, which itself is based on the 1982 French graphic novel *Le Transperceneige*. Set in a post-apocalyptic frozen earth, it follows the passengers of a perpetually moving train carrying the remnants of humanity. The extended periodic table is only seen briefly, but it includes a fictional g-block of elements underneath the lanthanide and actinides. In theory these elements would start with atomic number 121, since 119 and 120 would be under the alkali metals and alkaline earth metals. Finally, also worth a brief mention here is the *Stargate SG-1* TV series (1997–2007) episode *The Torment of Tantalus* (1997) in which the team discover a room with a holographic image of 146 elements. It is claimed that the elements act as a 'Rosetta Stone', *i.e.* a type of universal language.

Next, we move onto the tongue-in-cheek material 'unobtanium', which is a synonym to the illudium mentioned in *Duck Dodgers in the 24½th Century* (1953). It is theorised that 'unobtanium' was first coined by real-world aerospace engineers when referring to materials capable of withstanding the extreme heat of spacecraft re-entry. It was first mentioned in the 1950s but gained widespread prominence in the 1990s. One of the first onscreen appearances of the term is from the US sci-fi disaster film *The Core* (2003). It appears as the nickname for the "tungsten–titanium crystal alloy" that can absorb the extreme pressure and heat of the Earth's core, while also converting this into energy to power the vessel (*Virgil*). It also powers the laser-based high-speed drilling array which allows them to tunnel into the Earth so they can detonate nuclear bombs in the correct sequence, restarting the Earth's core.

Unobtanium also appears in *Avatar* (2009), which is a modernisation of *Pocahontas* (1995), itself a romanisation of the real story of Pocahontas. Both *Pocahontas* (1995) and *Avatar* (2009) deal with the same themes of colonialism and forbidden love. However, in *Avatar* (2009) the colonial humans are mining a rare mineral called unobtainium, instead of gold which is used as an excuse to annihilate the tribe in *Pocahontas* (1995). The unobtanium is found exclusively on the exomoon Pandora and is valuable because of its application as a powerful superconductor material. This is the reason why entire mountains with high concentrations of unobtanium levitate in the atmosphere of Pandora.

We then head over to the Marvel Cinematic Universe (MCU) to discuss the fictional wonder material adamantium, described as a "man-made alloy". The name is derived from the word 'adamant', from the Greek word *adamas*, meaning invincible. There are some references to adamantium[6] predating the Marvel comics, but it first appears onscreen in association with Wolverine from the X-Men. Adamantium is malleable when superheated and left in a liquid form, but as explained in *X2: X-Men United* (2003), it's virtually indestructible when cooled. It does however seem to have magnetic properties since it is susceptible to Magneto's abilities. The main opponent to adamantium in the Marvel canon is another fictional material called vibranium.

Vibranium first appeared in Marvel comics in 1966 in association with Daredevil. However, it doesn't appear onscreen until *Captain America: The First Avenger* (2011), in which Howard Stark states the element is stronger than steel, weighs one third as much, and is completely vibration-absorbent. Eventually Steve Rogers (Chris Evans) uses it to make a shield when he becomes Captain America. It is later used in spacecraft hulls and can also be used as a power source. It can be weakened by gamma radiation and individuals associated with gamma radiation like the Hulk.

The origin of the element on earth is explained in *Black Panther* (2018), which begins when a meteorite crashes into the continent of Africa containing "vibranium, the strongest substance in the universe". The vibranium is said to have affected the plant life around the meteor crash site. Five tribes settled on the area and called the country Wakanda, brought together by the first Black Panther who obtained his powers from a glowing plant. Black Panther's suit is made from vibranium-infused fibres which can "absorb kinetic energy" and redistribute it as a defence mechanism.[7]

The story references the real-world mining of coltan ore in the Democratic Republic of the Congo (DRC) which is a source of rare elements like niobium and tantalum. Previously, uranium was also mined in the same country so vibranium's use as a power source may also be a reference to that. The vibranium has made Wakanda a rich and technologically advanced country, hinting at the possibilities for African nations with valuable resources. It also references colonialism and raises the issue of ownership of ancient artifacts taken from African nations for display in foreign museums.

Sticking with Marvel, next is *Iron Man* (2008) in which Tony Stark (Robert Downey Jr) is injured from a blast which causes shrapnel to lodge in his system. While captured, a fellow captive and doctor creates an electromagnet which keeps the shrapnel from entering his heart. To power the electromagnet, Stark uses his father's fusion-type "arc reactor" incorporating a palladium core. He also uses the "arc reactor" to supply the energy needed for the Iron Man suit.

However, in *Iron Man 2* (2010), he begins to suffer "palladium poisoning" due to the composition of the arc reactor, which he treats with "lithium dioxide". Unfortunately, the chemistry is backwards: palladium metal is relatively inert and unreactive, hence it's use in jewellery. Lithium dioxide is unlikely to exist due to valency – it would only be stable at -230 °C. It would also rapidly revert to lithium oxide, which is toxic and reacts violently with water.

After searching through his father's old notes and video footage, he discovers a new element from his father's (Howard Stark) old model of the 1974 "Stark Expo" in which the buildings represent the number of protons and neutrons. He then builds a particle accelerator in his basement to create this new element with an atomic number of 204, which is way beyond the largest currently known element called oganesson (118). This allows him to create a new and more powerful arc reactor which also cures his palladium poisoning. The cameo appearance of Captain America's shield hints that this element is vibranium. Interestingly, there are several references to the 1950s atomic age associated with Howard Stark, complete with traditional atomic symbols.

Finally, there are two other examples that are worth a brief mention. First is the "protomolecule" in *The Expanse* (2015–2022), which is described as an extraterrestrial agent, but not necessarily a lifeform. It is often described as a flexible tool, capable of creating and functioning for a variety of purposes throughout the series. Everything it builds appears to be made from carbon and silicon. The second is from the South Korean miniseries *The Silent Sea* (2021), which follows a space crew tasked with retrieving samples from an abandoned research facility on the moon. They discover a new material called "lunar water", which appears like water but multiplies like a virus when it meets living cells. There are also several scenes in labs showing lab glassware such as funnels, conical flasks, and beakers.

10.7 SUPERHERO CHEMISTRY

In addition to fictional materials, there are also some superhero depictions with other examples of onscreen chemistry. We start with a return to deadly gases in *Spider-Man* (2002), when Norman Osborn becomes the Green Lantern (Willem Dafoe). He starts by ingesting a "gene-altering chemical" which is then activated by a green gas that gets pumped into a chamber. Worth noting is the famous line from Uncle Ben that could be another reference to the 'research-warning' trope: "with great power comes great responsibility". In the sequel *Spider-Man 2* (2004), the nuclear fusion scientist Dr Otto Octavius (Alfred Molina) becomes fused to his AI mechanical arms in an accident. The AI then becomes sentient and takes control, providing another example of a side effect from a scientific invention.

A year later we return to the *Dr Jekyll and Mr Hyde* story with the first feature film adaption of the *Hulk* (2003), which explores the origins of the transforming scientist. It opens in a biological research lab with beakers, test-tubes, conical flasks, and gel electrophoresis. There are also notes about "human regeneration" and manipulating the immune system using DNA from marine life. The scientist starts to self-experiment and begins to suspect that his mutated genes have been passed on to his child Bruce. Eventually, Bruce becomes a scientist at the "Berkely Nuclear Biotechnology Institute" and is accidentally exposed to gamma radiation which activates his dormant genes.

We then change perspective from the monster to the monster hunter in *Van Helsing* (2004). Like *Mad Monster Party?* (1967), the plot combines well-known characters from gothic horror stories like Frankenstein, Dracula, and werewolves. It opens with a black and white scene in Victor Frankenstein's lab located inside Dracula's castle, complete with lots of electrical sparks but not glassware. Unlike the previous depiction of Van Helsing in *Dracula* (1931), here he takes the form of an ancient James Bond action hero. There is a brief scene in a lab with glassware filled with colourful liquids as Van Helsing is assigned a mission to vanquish Dracula. A lab assistant (like Q from the *James Bond* franchise) supplies him with weapons, garlic, various gadgets, and demonstrates a new explosive liquid that he calls "glycerine 48". Also noteworthy, Van Helsing kills Dr Jekyll after a prolonged brawl with a giant, Hulk-like ultra-strong Mr Hyde in Paris.

We then have a reboot of the arachnid superhero with *The Amazing Spider-Man* (2012). The key scenes are when Peter Parker (Andrew Garfield) impersonates an intern to get into the Oscorp industrial labs to find information about his father. The labs are shown with glassware and instruments as Gwen (Emma Stone) provides a tour. Parker breaks away from the tour and enters a restricted lab where he gets bitten by a genetically modified spider. Worth noting, the lead scientist who worked with Peter's father explains that they were called 'mad scientists' for their work on "cross-species genetics". In a reference to *Spider-Man* (2002), Peter is also shown a device which can disperse gases efficiently: "load it with an antigen, it creates a cloud, it could be dispersed over a neighbourhood... but it could also be loaded with a toxin". Later, Gwen uses this device to create a "reptilian antidote".

Then, in a homage to *The Alligator People* (1959), Peter works with the scientist to use reptile DNA to help grow back limbs in mammals. The scientist then self-experiments in an attempt at growing back his missing arm using the same technology that made Peter into Spider-Man. As is tradition with onscreen self-experimentation, there is an unforeseen side effect which causes the scientist to transform into a large reptilian monster. Later, we briefly see the school chemistry lab when the monster hunts down Peter. The monster/scientist takes a conical flask containing a green liquid and mixes it with some yellow powder, which explodes when he throws it. Finally, we also have some 'cold chemistry' when Peter uses liquid nitrogen to help defeat the lizard.

Next, in *Spider-Man: Homecoming* (2017), we have an example of Peter (Tom Holland) doing a side experiment in a school lab with similarities to a scene in *October Sky* (1999) which will be discussed in the next season. Peter is wearing safety goggles while he investigates substances to use as a "web fluid". His notepad contains the correct chemical formulae for salicylic acid $C_7H_6O_3$, toluene $C_6H_5CH_3$ and methanol CH_3OH. However, the chemical structure shown has carbon atoms labelled with no hydrogens. Also, none of the substances on his pad would form a web fluid or even a polymer.

He proceeds to mix some yellow liquid into a beaker containing a white substance which expands and stretches like nylon. Due to this similarity, it is very possible that this is referencing the real polymerisation experiment in which nylon can be made from decanedioyl dichloride in cyclohexane floated on an aqueous solution of 1,6-diaminohexane.[8] In the background of this scene we can also hear the teacher make a science joke when he says "today we are going to talk about the Danish physicist Neils Bohr, but trust me, there's nothing boring about his discoveries".

We finish our discussion about the two-legged spider with *Spider-Man: No Way Home* (2021) with three versions of the hero working in a lab to create an anti-serum for the Green Lantern and the other villains. Of note, they all wear safety googles, but Andrew Garfield's version frequently has his goggles on top of his head or doesn't wear them. There are numerous chemical containers, glassware, fumehoods, and Bunsen burners around the lab. Ultimately, they are successful, and their chemistry problem solving skills save the day. Stories which involve the use of chemistry to solve problems will be discussed in more detail later.

Worth a brief mention here also are some scenes in *Hellboy* (2019) featuring various items of lab glassware such as a condenser and conical flasks. There are also some scenes in the TV series *The Flash* (2014–2023) that include colourful liquids in glassware. During the execution of the villain Kyle Nimbus, the dark matter from the S.T.A.R. Labs particle accelerator explosion causes his "molecular structure" to fuse with hydrogen cyanide.

Finally, we finish our discussion here with *Black Panther: Wakanda Forever* (2022), in which Princess Shuri (Letitia Wright) leads the scientific research. The movie opens with her brother, King T'Challa, suffering from an illness, which she attempts to cure by synthetically recreating a rare herb in the lab. Despite showing traditional lab glassware scattered on benches, the synthesis is performed by a "printer" in the technologically advanced lab. However, her brother dies before she has a chance to test her synthesis.

Like the first movie, Shuri developers new armour and weapons from vibranium to help defend the country from outside attack. Later, a US science student develops a vibranium detector as part of her metallurgy class which gets used by the CIA to discover a new source of the material under the Atlantic Ocean. When Shuri acquires the detector, she deconstructs it to figure out how it works, saying "this thing can detect the altered frequency of vibranium, through water, stone, even heavy metals". The two black scientists then work together to help defend Wakanda from the threat of the underwater kingdom of Talokan.

REFERENCES

1. E. Fu, A. Fitzpatrick, C. Connors, D. Clay, B. Toombs, A. Busby and C. O'Driscoll, Public Attitudes to Chemistry, *R. Soc. Chem.*, 2015, 16.
2. C. A. Frey, M. L. Mikasen and M. A. Griep, Put Some Movie Wow! In Your Chemistry Teaching, *J. Chem. Educ.*, 2012, **89**(9), 1138–1143.
3. T. J. Brumovska, S. Carroll, M. Javornicky and M. Grenon, Brainy, Crazy, Supernatural, Clumsy and Normal: Five Profiles of Children's Stereotypical and Non-Stereotypical Perceptions of Scientists in the Draw-A-Scientist-Test, *Int. J. Educ. Res. Open*, 2022, **3**, 100180.
4. P. Laszlo, One the Self-Image of Chemists, 1950–3000, *HYLE – Int. J. Philos. Chem.*, 2006, **12**(1), 99–130.
5. A. C. Hoffman, The Health Effects of Menthol Cigarettes as Compared to Non-Menthol Cigarettes, *Tob. Induc. Dis.*, 2011, **9**(Suppl. 1), 1–9.
6. M. Jameson, Devil's Powder, *Astounding Sci. Fict.*, 1941, **6**(4), 69–78.
7. S. N. Collins and L. Appleby, Black Panther, Vibranium, and the Periodic Table, *J. Chem. Educ.*, 2018, **95**(7), 1243–1244.
8. Royal Society of Chemistry, *Making nylon: the 'nylon rope trick,'* RSC Education, https://edu.rsc.org/experiments/making-nylon-the-nylon-rope-trick/755.article.

Chemistry in Education

11.1 COLLEGE CHEMISTRY

So far, university labs have been no stranger in onscreen chemistry. Research-focussed labs featured in *It Happens Every Spring* (1949) and *Real Genius* (1985) among others, and we also saw a teaching-focussed lab in *The Affairs of Dobie Gillis* (1953). However, from the early 1990s onwards, schools and colleges began to dominate the big and small screen, giving us well-known classics like *Dangerous Minds* (1995). Eventually this brought with it a whole host of new university and college labs featuring onscreen chemistry, which will be discussed in this section. It also resulted in rapid expansions in school labs onscreen as well, which will be discussed in the next section.

We kick off this discussion with a remake of *The Nutty Professor* (1996), once again referencing the *Dr Jekyll and Mr Hyde* story. However, this time the 'formula' turns the overweight Sherman Klump (Eddie Murphy) into the thin, muscular, and confident Buddy Love. Like previous versions of the story, the effects of the weight-loss formula are temporary and using it becomes additive. There is also a conflict between Klump and Love, emphasising the link to *Dr Jekyll and Mr Hyde*. The lab includes dozens of hamsters for testing in addition to conical

Onscreen Chemistry: The Portrayal of Chemical Science in Film and TV
By John O'Donoghue
© John O'Donoghue 2025
Published by the Royal Society of Chemistry, www.rsc.org

flasks with magnetic stirrers filled with colourful liquids and connected with tubing.

Later, we also see large molecular models on his desk in the lecture room, but it's not clear what they represent. After one of the hamsters loses weight, the researchers conclude that their formula has "rearranged its DNA". Of note, just like in the previous depiction of *The Nutty Professor* (1963) as well as *The Absent-Minded Professor* (1961), there is a subplot about the need for philanthropic money to fund science at the college. Also, Klump's love interest here identifies as a chemistry graduate (Jada Pinkett Smith) who tells Klump that she is "teaching my first chemistry intro class across the hall".

The transformation scene takes place in a college lab similar to the 1960s version of the movie, *i.e.* not a home lab like the gothic originals. Once again, the transformation scene includes an elaborate glassware apparatus, but with a computer added this time. The weight-loss formula is finished when Klump uses a dropper to add a colourless liquid to a small volumetric flask of colourless liquid, which immediately begins to glow blue. This is most likely a reaction of luminol with a drop of hydrogen peroxide. This is one of the best-known examples of chemiluminescence and luminol is also used as an investigative tool at crime scenes to reveal hidden remains of blood.

The popularity of the remake also led to the sequel *Nutty Professor II: The Klumps* (2000). In a throwback to *Monkey Business* (1952), this time we have a de-aging formula. However, Klump's thinner alter ego from the first movie (Buddy Love) is still trying to take control of him without the formula. Klump also sees Buddy in mirrors instead of his own reflection, referencing the spontaneous transformations from the gothic version of the story, as well as the use of mirrors as previously discussed. Like before, some glassware, chemical containers and instrumentation can be seen in the lab. The DNA researcher (Janet Jackson) also displays some fictional molecular structures on a projector screen during her lecture. Klump removes Buddy from his system by cutting out "part of his DNA". However, the resulting gloop later reforms with some dog hair to bring Buddy to life as a separate person.

Next, we temporally leave the world of fictional chemistry and return to subtle onscreen chemistry references in *Good Will*

Hunting (1997). This tells the story of a janitor (Matt Damon) whose mathematical genius and eidetic memory is accidentally discovered by a professor at MIT (Stellan Skarsgård). However, he has a troubled past and the mathematician arranges for him to see a therapist (Robin Williams) at a community college. Later, we also meet Skyler (Minne Driver) who is studying medicine at Harvard. Her computer screensaver is a 3D molecule that appears to be an imidazole. These are used in the real world as antifungal drugs, and it is shown while Skyler makes fun of "open-toe sandals".

In the same scene, she also tells Will that she can't go on a date with him because she needs to "assign the proton spectrum for ibogamine – although that sounds really interesting, it's actually fantastically boring". To speed the process along, Will heads off to the park where he scribbles down the molecular structure on a piece of paper along with his proposed ^{1}H NMR (nuclear magnetic resonance) assignments. It appears that the structure of ibogamine he uses is correct, but we don't see his NMR assignments. Ibogamine is a real-world anti-convulsant and represents a reasonably tough problem for ^{1}H NMR assignments. (Figure 11.1)

Later, during a conversation about organic chemistry, Skyler states that "nobody studies it for fun". This further implies that Skyler is only studying chemistry because it is a requirement for medicine, which is true in the real world as mentioned previously in relation to *Dead Poets Society* (1989). Overall, chemistry is grouped with maths throughout the movie as subjects that only a genius can excel in. Will says his ability to understand organic chemistry comes naturally to him, like Beethoven playing piano, whereas Skyler states that she must work hard at it.

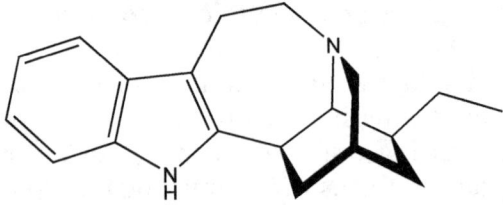

Figure 11.1   The structure of ibogamine as seen in *Good Will Hunting* (1997), where it is shown briefly as a hand-drawn structure for a ^{1}H NMR assignment problem. © Be Gentlemen, Miramax Films.

However, noteworthy is the fact that Skyler is accepted into the Stanford Medical School, proving that hard work can be as successful as natural genius.

We then stick with organic chemistry for some scenes in the teaching lab in *The Prince and Me* (2004), the first college chemistry teaching lab onscreen since *The Affairs of Dobie Gillis* (1953). The lab contains some molecular models, rotary evaporators, beakers, and other glassware. The key chemistry scene takes place when Eddie (Luke Mably) doesn't show up to a lab practical, leaving his lab partner Morgan (Julia Stiles) to do the experiment on her own. While wearing a lab coat and safety goggles, she sets up a three-necked round-bottomed flask with a condenser, thermometer and a separatory funnel, into which she adds hydrochloric acid as instructed. However, she then adds the acid from the separatory funnel to the flask too quickly, causing the reaction to foam up and squirt out the side of the flask. It's a pity that the experiment isn't explained because this is otherwise an accurate teaching lab scene, since this can really happen if a reagent is added too quickly. Note that Morgan is taking the class because she is studying medicine.

Next, we move from universities to community colleges in the US sci-fi comedy *Evolution* (2001), about nitrogen-based alien lifeforms which arrive on Earth on a meteor. There are real theories that nitrogen could mimic carbon by forming long-chain molecules in very low temperature environments, *i.e.* like that found in space.[1] However, in reality, nitrogen-based life forms are very improbable for a variety of reasons. Carbon is a highly versatile and unique element, making it the cornerstone of biological molecules. It should be noted, though, that nitrogen is already a large part of biochemistry, in partnership with carbon and hydrogen.

The key onscreen chemistry scene takes place in a college teaching lab, where they realise that fire will make the alien creatures evolve even faster. While they work together to come up with a solution, the protagonist (David Duchovny) spots a pattern on the periodic table on Allison's (Julianne Moore) shirt. He then links the relationship between carbon-based lifeforms and "our poison" arsenic to nitrogen-based lifeforms and their potential poison selenium. As they debate where to get large quantities of selenium, his student declares that the active ingredient in anti-dandruff shampoo is "selenium sulfide", which is true for the clinical-strength versions. *Evolution* (2001) is therefore a good example of teamwork and problem solving, themes which will be discussed in further detail later.

We now return to TV land with *Greek* (2007–2011), about fraternities at a fictional US university. The story follows a quintessential nerd undertaking the university engineering programme who is occasionally shown in a lab, wearing a lab coat, and surrounded by glassware. However, the lab scenes mainly serve as a background for character development rather than showcasing any lab work. They also serve to contrast with the fraternities which take up most of the plot. The fraternities and the labs represent the internal conflict of the protagonist since he wants to take his studies seriously but also wants to fit in.

Finally, we have the US TV sitcom *The Big Bang Theory* (2007–2019), which centres around a group of friends that includes several research scientists. From a scientific perspective, the characters include an experimental physicist, a theoretical physicist, an aerospace engineer, a microbiologist, an astrophysicist, and a neurobiologist. However, chemistry only appears as brief references throughout. Notably, however, the real chemistry Nobel Prize winner Francis H. Arnold makes a brief appearance in the episode *The Laureate Accumulation* (2019) to support Sheldon.

Other chemistry references include numerous images of the periodic table, which include Leonard's t-shirts and the shower curtain in the apartment bathroom. There are also chemical elements on the white board in the apartment sometimes in addition to accurate chemical equations. We also see various chemical lasers, such as a "500 kW oxygen–iodine laser" used by Leslie to heat up her noodles. Also, in the episode *Romantic Resonance* (2013) Sheldon claims to have discovered a new stable super-heavy element. Shown on the whiteboard are the correct shorthand electron configurations of some heavy elements such as livermorium ($[Rn]5f^{14}6d^{10}7s^27p^4$).

There is also a real chemistry experiment depicted onscreen in *The Vengeance Formulation* (2009) in the form of a well-known demonstration called 'Elephant's Toothpaste'. He carries out the reaction in the kitchen, adding a liquid from a conical flask into a measuring cylinder. There are also four other random flasks containing colourful liquids on the counter for visual effect. This real reaction involves the rapid decomposition of hydrogen peroxide catalysed by potassium iodide in the presence of soap. However, onscreen this is explained as "combine these chemicals with ordinary dish soap, creating a little exothermic release of oxygen". The chemicals are not named, and he isn't wearing a lab coat, but at least Sheldon is wearing safety goggles and gloves.

For the most part, the science depicted in *The Big Bang Theory* (2007–2019) is accurate, with acceptable exaggeration used from time to time for comedic or visual effect. For instance, the oxygen–iodine laser mentioned above is a near-infrared laser and the laser beam shown on screen is red. Although we shouldn't be able to see this with our eyes, the colour is on the correct end of the spectrum, unlike in *Real Genius* (1985). In the background there are also hundreds of accurate scientific references too numerous to mention individually, thanks to their consultant physicist Prof. David Saltzberg.

However, it must be said that the long-running show also reinforces some outdated stereotypes about scientists and frequently encourages laughter at nerd culture and their associated behaviours. Most of the episodes are written with the aim of mocking Sheldon's flaws and his difficulties navigating social situations. Frequently, it also depicts unprofessional conduct and confrontational competition in relation to scientific discoveries. However, it also portrays female scientists in lead roles. It also stands apart for depicting scientists of all genders with interests and lives outside of the lab. This contrasts with the many examples of 'parachute scientists' who only appear for a single scene, which will be discussed later for *Erin Brockovich* (2000) and *Dark Waters* (2019) among many others.

11.2 BACK TO SCHOOL

So far, we've only seen a handful of school labs onscreen and most represented negative portrayals of chemistry and teachers. In the previous season we discussed a biology lab with an evil biology teacher in *The X-Files* (1993–2002). Around the same time, on the other side of the pond, we have an episode of the British slapstick comedy *Mr Bean* called *Back to School Mr Bean*

(1994). In the episode, Mr Bean (Rowan Atkinson) attends the open day of a school where the usual chaos ensues across several classes, including "Chemistry Lab 2".

The lab opens with an elaborate glassware apparatus with random tubing in a style that has now become synonymous with onscreen depictions of chemistry. From left to right there is a separation funnel joined to a conical flask on a tripod filled with a blue liquid, joined to a condenser, joined to another conical flask, joined to some round-bottomed reaction vessels and then into a bucket. There is water pumping through the apparatus from right to left, *i.e.* backwards through the condensers. There is also a beaker with some dry ice bubbling and a bottle with a skull symbol on it.

Bean picks up a beaker containing a dark green liquid and adds it to a beaker of colourless liquid which immediately turns red. He then adds the same green substance into the bucket which turns all the liquids in the system blue. This is most likely universal indicator which is used for acids and bases. The apparatus begins to shake and emit smoke, at which point Mr Bean runs away while a student enters the lab. Off screen there is an explosion resulting in a large cloud of blue smoke. Neither of them are wearing safety goggles or a lab coat, and later the student is seen again covered head to toe in a blue powder.

Next, we have the US sci-fi horror *The Faculty* (1998), about extraterrestrial parasites who take control of the teachers. There is a brief look at the school lab when one of the students brings the parasite to the science teacher for examination. The lab is biological in nature, with microscopes and beakers on every bench. The movie also opens with references to recreational drugs, which are distributed inside ball-point pens. However, it is later discovered by accident that the homemade drugs kill the parasites.

The drug maker also has a home lab with a complicated setup of glassware joined together by tubing, which is coupled to the sound of bubbling. There are numerous condensers, round-bottomed flasks, funnels and measuring cylinders present. However, the liquids are not overly colourful, mainly yellow. It is then revealed that the drug is just powdered caffeine and is effective against the aliens because it dehydrates the host. Later, one of the infected students destroys the entire apparatus, which has become a popular onscreen activity seen previously in *Never Say Never Again* (1983) and *Batman and Robin* (1997).

Following the previous discussion about *The Insider* (1999), in the same year we also have several other examples of school labs on screen which we can divide into two general categories: (a) scenes that are relevant to the plot, like in *The Insider* (1999), and (b) scenes with no connection to the plot. An example of the latter is the US teen romantic comedy *10 Things I Hate About You* (1999), about a new student (Heath Ledger) who falls in love with another student (Julia Stiles) whose father is restrictive of dating.

The science lab is shown with some glassware in the background while the students are dissecting a frog, but the science teacher is never seen or heard. The only discussion is about a plan to set up a female student with the new student and the only activity is when Heath Ledger's character lights a cigarette from a Bunsen burner. Similarly, there is also a scene in *Election* (1999), about student elections in a US high school. Briefly, we are shown Reese Witherspoon's character wearing safety googles and watching a blue liquid boil in a beaker over a Bunsen burner, but nothing is said about it and we never seen it again.

In the same vein we also have *Juno* (2007), which features some sodium chloride crystal lattice molecules in the background of a school lab. The technicoloured throwback *Hairspray* (2007) also features a brief scene in a school lab with a conical flask boiling over a Bunsen burner. There are also colourful test-tubes and a student using a dropper, all while we hear liquids bubbling. There are also brief scenes in school biology labs with beakers in *Jennifer's Body* (2009), *Ender's Game* (2013) and *When You Finish Saving the World* (2022). In all these examples, the use of a school science lab is not connected to the plot in any tangible way. However, sometimes there are subtle connections like in *Tortilla Soup* (2001), in which the discipline of chemistry is linked with romantic chemistry, and in *The Tomorrow War* (2021), which establishes the protagonist's biology background.

All these examples contrast with the US drama *October Sky* (1999), which features a school lab scene that is important to the plot. The story is based on the 1998 memoir *Rocket Boys* by Homer Hickam, with the movie title being an anagram of the book title. It tells the true story of Hickam (Jake Gyllenhaal), a coal miner's son, who is inspired into rocketry by the launch of *Sputnik 1* in 1957. The movie opens in the coal mine with everyone hunched over and closes with everyone looking up to the sky as they watch his rocket launch.

The science teacher (Laura Dern) is portrayed as encouraging and proud, and helps the boys in any way she can. In the real world, the Freida J. Riley Teacher Award recognizes an American teacher who overcomes adversity or makes an enormous sacrifice to positively impact students. It is given in honour of Ms Riley, the science and maths teacher who inspired the boys and taught for most of her career while suffering from Hodgkin's disease.

The scene in the school lab involves the boys trying to find a fuel for their rocket. They are conducting what seems to be a lab practical exam and the substitute teacher tells them "Ms Riley wanted to make sure you document your results". On the chalkboard there is a list of solvents such as hexane and propanol along with their correct densities and boiling points. The boys are conducting their own side experiment, stating "potassium chlorate has a potassium atom; we mix it with sugar, add heat, it gives three parts oxygen and two parts carbon dioxide". This is a real chemical reaction, often called 'The Screaming Jelly Baby' in terms of demonstrations. On screen it provides a purple flame, which is also accurate.

However, as the teacher starts to walk towards them, they dump the mixture down the drain. Elsewhere, another student throws a lit splint into another drain which causes a burst of flame from all the drains. This reaction is greatly exaggerated and highly unlikely, further reinforcing the stereotype of 'chemistry causes explosions'. However, it should also be said that here this might be excused since this was the purpose of their research, *i.e.* they want an explosion to launch a rocket.

Similarly, we also have *High School Musical* (2006), which features a 'Chem Club' involved in a Scholastic Decathlon competition. The school lab is shown with taps, large containers, and glassware. Later the correct chemical structure for benzoic acid is on the chalkboard and a periodic table can be seen in the background. After impressing with her mathematical knowledge, the new student Gabriella (Vanessa Hudgens) also demonstrates her talent for chemistry. The other members of the Chem Club try to recruit her by telling her to follow in the footsteps of Marie Curie and others.

However, after some initial references to conflict and in-compatibility between science, art and sports, the two prot-agonists decide to juggle their commitments. Gabriella joins the Chem Club and the musical, while Troy (Zac Efron) con-tinues with the basketball team and also joins the musical. Later, Gabriella correctly balances the chemical equation for the reaction of zinc with nitric acid. We also see the students, wearing safety goggles, conduct an acid–base reaction which fizzes in a beaker (most likely vinegar and baking soda).

At the chemistry competition the students are wearing lab coats as they solve chemical equations on whiteboards. All the chemical equations are related to CO_2 and are correct, as well as the figures for their enthalpy of formation, *e.g.* $C_{(s)} + O_{2(g)} \rightarrow CO_{2(g)}$, $\Delta H_f = -394$ kJ mol^{-1}. The Chem Club then make it pos-sible for Gabriella and Troy to attend the musical rehearsals by remotely cutting the power to the basketball game and setting up a conical flask to boil over, emitting a strong smell. Gabriella and Troy start their duet in a tracksuit and lab coat, respectively, but eventually she rids herself of the lab coat before the end of the performance. Presumably this represents her decision to leave chemistry behind and explains why there is no onscreen chem-istry in either of the two sequels.

11.3 MAKING AND TESTING DRUGS

We start this discussion with an example of 'breaking bad' on the big screen in *The 51st State/Formula 51* (2001), about a pharmacologist (Samual Jackson) who loses his license to practice. Later, he creates a new wonder drug (POS 51) from "over-the-counter ingredients", but it turns out to be a placebo designed to fool analytical tests into looking like a potent drug. The formula is partially shown later using heat to reveal in-visible ink, as previously seen in *The Name of the Rose* (1986). However, the chemical structure contains an impossible dou-ble-bonded hydrogen. He is referred to as the "Master Chemist" throughout and one of the labs features pipettes, fumehoods, and rotary evaporators. He also uses some non-descript pink liquid to blow up the lab to escape from his gangster boss (Meat Loaf) so he can sell the pills in the UK.

After the first drug-dealer chemist gets killed due to a misunderstanding, the second one uses test-tubes, droppers and chemicals to test the new tablet while the protagonist explains "MDMA utilises serotonin, opiates like heroin utilise dopamine, amphetamines increase adrenaline, cocaine gets those synapses in the brain firing really fast, my product is 51 times stronger than cocaine, 51 times more hallucinogenic than acid, and 51 times more explosive than ecstasy". The chemist displays some pink liquids in test-tubes which apparently confirms his story from an analytical viewpoint. Later, another drug dealer states "what is chemistry but the ability to attract adoration in others?".

In the same year, worth a brief mention also is *Blow* (2001), in which a character uses a melting point (MP) apparatus to determine the MP of a cocaine sample as 187 °F (98 °C). He states that the MP of pure cocaine is between 185–190 degrees (the real MP is 208 °F) and the closer a sample is to this number, the purer it is. This is an interesting throwback to a similar scene with heroin previously discussed in *The French Connection* (1971).

We then return to the school lab for the popular TV series *Breaking Bad* (2008–2013). Previously, we've only met two other examples of 'breaking bad' in a school setting. The first was *The Belles of St. Trinian's* (1954), in which the students were making gin in the school chemistry lab and selling it. The second was *The Faculty* (1998), which featured a student selling drugs in the school, albeit he was making them in a home lab. *Breaking Bad* (2008–2013) therefore represents an evolution of the theme, but its popularity greatly contributed to a new image for chemists and teachers.[2]

The plot of *Breaking Bad* (2008–2013) follows a US high-school chemistry teacher, Walter (Bryan Cranston), who discovers that he has terminal cancer. To provide for his family and pay for the treatment bills, he decides to use his chemistry knowledge to make illegal drugs. He teams up with a local drug dealer and former student, Jesse (Aaron Paul), to distribute the homemade methamphetamine. However, the onscreen chemistry doesn't just revolve around drugs. There are also scenes in the school chemistry lab where Walter is explaining real chemistry concepts with accurate chemical equations on the chalkboard. Also, Walter frequently uses his chemistry knowledge to get out of sticky situations, although most of the onscreen chemistry is concentrated in the early seasons.

In the *Pilot* (2008) episode, he poisons two rival gangsters by adding red phosphorus to a pot of boiling water which produces a lot of sparks. He explains to Jesse "red phosphorus in the presence of moisture and accelerated by heat yields phosphorus hydride, phosphine gas, one good whiff and...". The chemistry is accurate, but it should be noted that the standard industrial process for making phosphine uses white phosphorus, not red, as well as sodium hydroxide: $P_4 + 3NaOH + 3H_2O \rightarrow 3NaH_2PO_2 + PH_3$. Although they both have the same chemical formula, red and white phosphorus are allotropes, like carbon existing as graphite or diamond. White phosphorus is more reactive than red.[3]

Following this, Walter and Jesse need to dispose of the bodies in *Cat's in the Bag* (2008), so Walter proposes to use a bath of hydrofluoric acid (HF). He obtains the acid from his high-school chemical stories, but it is very unlikely that any school would be stocking HF. Walter explains to Jesse that HF will react with metals, certain plastics, and glass, which is why it is stored in Teflon (PTFE) bottles. However, Jesse cannot find anything big enough for the bodies, so he decides to use a bathtub. This results in the acid and bodies dissolving through the tub and the floorboards.[4] This scene was later 'busted' by *MythBusters* on their *Breaking Bad Special* (2013) when they immersed samples of ceramic, wood, plaster, metal, and raw pork (to simulate human flesh) in hydrofluoric acid. The acid did break down the ceramic and the plaster, but the other samples showed only mild corrosion.

In terms of making the methamphetamine, they use thermite to break the lock into an industrial warehouse to obtain the methylamine starting material in *A No-Rough-Stuff-Type Deal* (2008). Walt correctly explains the reaction between iron oxide and aluminium power used for the thermite.[5] Later, Walt and Jesse pose as fumigators to set up their new mobile lab in a house in *Hazard Pay* (2012). During the lab montage scene we see some instruments like a gas chromatograph, as well as real-world chemical insecticides like sulfuryl fluoride and cyfluthrin. They use large reaction vessels, add a metal catalyst, test the pH using a dropper and beaker, and extract the gases to the outside. During a break in production, Walt and Jesse sit on the couch drinking beer, a scene which is later used as a 'couch gag' in *The Simpsons* (1989–present). Their yellow hazmat outfits will also be discussed again later.

Finally, in anticipation of a drug deal going sour, Walter switches the crystal meth for crystals of mercury fulminate (fulminate of mercury, as Walt calls it). He throws a crystal on the ground to cause a large explosion, allowing him to escape. *MythBusters* also tested this on their *Breaking Bad Special* (2013) and found that although the fulminate is shock sensitive and explosive, it is not as sensitive or as explosive as shown on screen. So, like much of the onscreen depictions of chemistry in *Breaking Bad* (2008–2013), this represents another exaggerated use of real chemistry.[6]

11.4   TRUE STORIES

We finished the previous season with the fictionalised true story *The Insider* (1999), which combined the themes of education, evil industry, and conspiracies. In the same year we also had the true story of *October Sky* (1999), discussed here earlier. However, there are also several other true stories containing onscreen chemistry during this era. The first is the US drama *Awakenings* (1990), about a research physician (Robin Williams) who works with catatonic patients who survived the 1919–1930 epidemic of encephalitis lethargica (sleeping sickness). Robin Williams's character is based on the real-life British neurologist Oliver Sacks.

At a conference, the protagonist learns about the discovery of the real drug ʟ-DOPA (Figure 11.2) and its success with

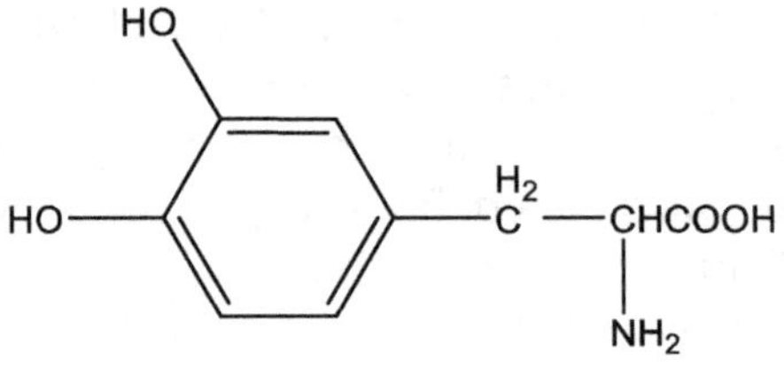

Figure 11.2   The correct structure of ʟ-DOPA reproduced from *Awakenings* (1990), where it is shown briefly on a projection screen at a conference. © Lasker/Parkes Productions, Columbia Pictures.

Parkinson's disease. When he asks the presenter about whether he thinks the drug would work for his catatonic patients, the presenter responds "I'm just a chemist, doctor, you're the physician. I leave it to you to do the damage". His colleagues express caution, comparing the drug to Freud's use of cocaine and the side effects associated with cortisone. His colleague states "let the chemist do the damage, doctor", referencing the real-world lack of coordination sometimes between those who make drugs and those who administer them.

Despite this, Sayer conducts a trial on one patient with astounding results. This leads to several scenes in the hospital lab, where chemical containers can be seen on the shelves in the background, as well as some glassware. Later, we see a correct periodic table of elements from this era and the protagonist claims he can "date his introduction to science by that". He goes on to correctly explain that the alkaline metals are on the left, with the halogens and inert gases on the other side.

In a similar vein, we have the US drama *Lorenzo's Oil* (1992), about dedicated parents (Nick Nolte and Susan Sarandon) who push scientists and doctors for a cure for their son's adrenoleukodystrophy (ALD), eventually leading to the development of 'Lorenzo's Oil'. The chemistry starts after they sponsor an international symposium of scientists doing research on ALD. One scientist mentions his studies about the addition of oleic acid to cultured cells, which lowered the quantity of long-chain saturated fats (C24 and C26). He describes oleic acid as "mono-unsaturated C18", which is correct since the chemical formula is $C_{18}H_{34}O_2$. He goes on to explain that pure oleic acid is toxic, it needs to be converted into the triglyceride form for human consumption.

This leads them to look for a producer of the triglyceride form of oleic acid, *i.e.* separating it from the saturated fats C24 and C26 in olive oil. They are eventually successful and receive some samples which show partial success. This is also known as 'triolein' in the real world and is one of two components of the real 'Lorenzo's Oil'. The other component is the triacylglycerol form of erucic acid obtained from rapeseed oil, which is mentioned in the movie as C22. The parents contact several chemical firms to obtain the erucic acid, which provides some scenes of labs and glassware. However, it is explained that removing the

C24 and C26 from rapeseed oil is very difficult and it would take the "best chemist over a year".

Eventually, they find an elderly British chemist who takes on the challenge of producing the oil during the last six months of his career, before his retires. He becomes obsessed with completing the challenge and is shown working on the distillation using a variety of glassware. Throughout the movie, scientific research is presented as slow and cumbersome, with the scientists and doctors insisting on doing proper studies to get FDA approval. The combined oil is ultimately successful in slowing Lorenzo's ALD, helping him live to the age of thirty in the real world. Since the movie was released, evidence has emerged to suggest that it does have a preventive effect when used by asymptomatic patients.[7]

Next, we return to space exploration in *Apollo 13* (1995), based on a true story which serves as the origin of the famous line "Houston, we've had a problem". The problem in this case is the rupturing of an oxygen tank, which forces the crew to return to earth rather than continue on to the moon. This leaves the crew with a vastly reduced clean air supply. They are forced to move out of the Control Module (CM) and into the Lunar Module (LM), but this was never designed to maintain three crew members for four days.

In the movie, the NASA scientists work around the clock to come up with a solution to the rising carbon dioxide (CO_2) problem in the LM. However, the LM uses a different format of CO_2 scrubber compared to the one in the CM. The crew therefore need to retrofit the "lithium hydroxide canisters" from the CM to work in the LM. The key scene involves the ground team explaining the process of retrofitting the canisters, but nothing more is mentioned about the chemistry. Lithium hydroxide does removes excess CO_2 by forming lithium carbonate by the reaction $2LiOH + CO_2 \rightarrow Li_2CO_3 + H_2O$.[8,9] This scene contrasts with *The Martian* (2015), another space movie with chemistry problem solving which will be discussed later.

In the same year we also have the British black comedy *The Young Poisoner's Handbook* (1995), a fictionalised version of the true story of Graham Young. The movie opens with the protagonist (Hugh O'Connor) expressing his fascination for chemistry and that he has nothing in common with his parents or siblings. He obtains some colourful chemical powders from a chemistry supply shop, but there are also several references to household chemicals like hair-removal products and cleaners such as ammonia. His bedroom lab is shown containing dozens of chemical containers, conical flasks, condensers, tubing, a funnel, and other items. Later, his stepmother calls him her "little Louis Pasteur".

Later, he goes in search of medical books for his "scientific research", one of which is the famous *Gray's Anatomy*. In one book he discovers antimony sulfide, which he describes as "one of nature's most volatile substances". He proceeds to heat it in a conical flask over a Bunsen burner, but he really should be using a boiling flask. Erlenmeyer (conical) flasks are not designed for constant high temperature. The output flask explodes in a cloud of sparks and smoke, failing to produce a crystal. Instead, he learns that antimony is also a poison, so he adds it to some mustard to hide the "acidic taste" and gives the sandwich to a classmate. This causes his classmate to be sick, allowing Graham to go on the date with the girl instead.

He then poisons his stepmother using antimony in response for the punishment she dished out on him earlier. He also uses a dropper of an unknown liquid to add something into his mother's medicine to keep her in a weakened state. He then discovers thallium, a toxic chemical element which he describes as "tasteless, colourless, odourless and untraceable". This causes her hair to fall out, severe constipation, and eventually death. Note that this concept is reversed a few years later in the *Sixth Sense* (1999), where a mother is caught adding an unknown substance to her daughter's medicine to keep her sick, eventually leading to her daughter's death.

Graham is convicted of murder and locked in an asylum. Later, the psychologist provides him with an elaborate chemistry apparatus to repeat his experiment with antinomy sulfide. The glassware contains colourful liquids, condensate, and bubbling solutions which allow him to create a crystal. Once he gets out of the asylum, he meets a colleague who is doing a PhD in "structural inorganic chemistry". He shows him around the photo lab, where they are using thallium in their camera lenses. Thallium oxide is used in the real world to produce glass with a high refractive index. The presence of the thallium causes him to revert to his old ways, eventually poisoning his co-workers.[10]

Next, we have the biographical legal drama *Erin Brockovich* (2000) and her case against Pacific Gas & Electric Company (PG&E) involving the contamination of groundwater with hexavalent chromium. In terms of the chemistry, this is a perfect example of a 'parachute scientist', who only appears for a single scene to explain the chemistry and is never seen again. We will meet some more parachute scientists later, with a similar example provided in *Dark Waters* (2019).

Unfortunately, our scientist here makes an error in his explanation of the different types of chromium. He claims that "straight up" chromium does "all kinds of good things for the body", while chromium(III) is "fairly benign" and chromium(VI) or 'hexavalent chromium' is very harmful, depending on the amounts. By "straight up" we are assuming he means chromium(0) metal, in which case he seems to be mixing this up with trivalent chromium(III). Trivalent chromium is naturally occurring and is the most chemically stable form of chromium. It is also present in food and supplements where it plays a role in regulating sugar and fat metabolism. However, he is correct in relation to hexavalent chromium, as this can be quite harmful if ingested.

Next, we finish with a similar story to that of *Lorenzo's Oil* (1992) in *Extraordinary Measures* (2010), which tells the true story of parents who work with a researcher to develop an enzyme to save the lives of their children. The eccentric researcher (Harrison Ford) is doing innovative research in enzymes with a "biological marker, mannose 6-phosphate," for the life-threatening genetic disease called Pompe disease. Interestingly, we return to the issue of scientific funding with a throwback to a similar line from *It Happens Every Spring* (1949), when he tells the parents that the "University of Nebraska pays their football coach more money in a year than my entire science budget". This prompts the parents to raise the money to fund the research and run the clinical trials, eventually leading to the formation of a new biotechnology company, However, the entire process exposes the differences between 'theory' and 'reality', in addition to the differences between businesses and science.

After an inspiring talk from the researcher, the venture capitalists quiz him on the details which expose some of the barriers that need to be overcome in order to bring something from a lab to market. After hiring a team of young scientists, he explains that "scientists get all sensible and careful when they get old; young ones like risk, and you can pay them less". The friction between the researchers and the parents who are from a business background is prominent throughout. Later, there are several scenes in different labs with lots of colourful liquids in a variety of lab glassware, and some lab skills are shown like using a micropipette.

11.5 YOUNG CHEMISTS

Previously we discussed several examples of chemistry in animations which mostly reflected similar themes to live-action depictions. However, the level of chemistry was scant and in most cases the scientists were openly referred to as 'mad'. One of the first transitions away from this stereotypical trope appears in *Big Hero 6* (2014), based on the Marvel superhero team of the same name. It follows a robotics prodigy called Hiro who takes part in illegal underground robot fights. The first robot fight opens with a reference to *Mad Max Beyond Thunderdome* (1985) when it is announced "two bots enter, one bot leaves".

Later, his brother brings him to his research lab at the robotics institute to convince Hiro away from bot fighting. The lab features a range of researchers, some of which are ultra-neat, while others are spontaneous and messy. The eccentric female chemist rolls in with a 400 lb ball of tungsten carbide. She then begins the experiment by opening the tap on her elaborate glassware apparatus, which, of course, is connected together with tubing. Beginning with the green liquid, she works her way along to the blue, then red, then yellow as she explains "a dash of perchloric acid, a pinch of cobalt, a hint of hydrogen peroxide, superheated to 500 kelvin". This creates a pink spray which vaporises the large ball of tungsten carbide into a cloud of pink dust in a process she calls "chemical metal embrittlement". In traditional fashion, there is also a bubbling sound throughout this scene.

The college mascot then joins them and explains that he is trying to convince the chemist to create a formula that will turn him into a "fire breathing lizard at will, but she says that's not science". He then suggests shrinking rays, invisible sandwiches, and laser eyes, all of which the researchers deny being "real science". This scene serves two purposes, first to ref. 1950s/1960s sci-fi movies and, second, it establishes *Big Hero 6* (2014) as based in the real world with accurate science. Immediately following this scene, his brother introduces Hiro to his inflatable medical robot Baymax, which is powered by a lithium-ion battery. The robot suggests that he uses an antibacterial spray, which contains the real antibiotic bacitracin and then displays the correct chemical structure on his stomach (Figure 11.3). Later, there is a fire in the lab, but we eventually discover that it was caused by the professor so he can steal Hiro's microbot invention.

From there we move on to the buddy cop action comedy *Zootropolis/Zootopia* (2016), set in a world filled with anthropomorphic animals. The pair are tasked with investigating why some of the animals have turned "savage" and discover that it is caused by "night howlers". Later, they learn that night howlers are the fictional "midnicampum" flowers which cause a severe psychotropic effect. They are referred to as a "Class C Botanical", implying that they are a controlled substance, and they are used as a pesticide.

Later they follow a ram named Doug to a secret laboratory in an old subway station where we see the flowers growing

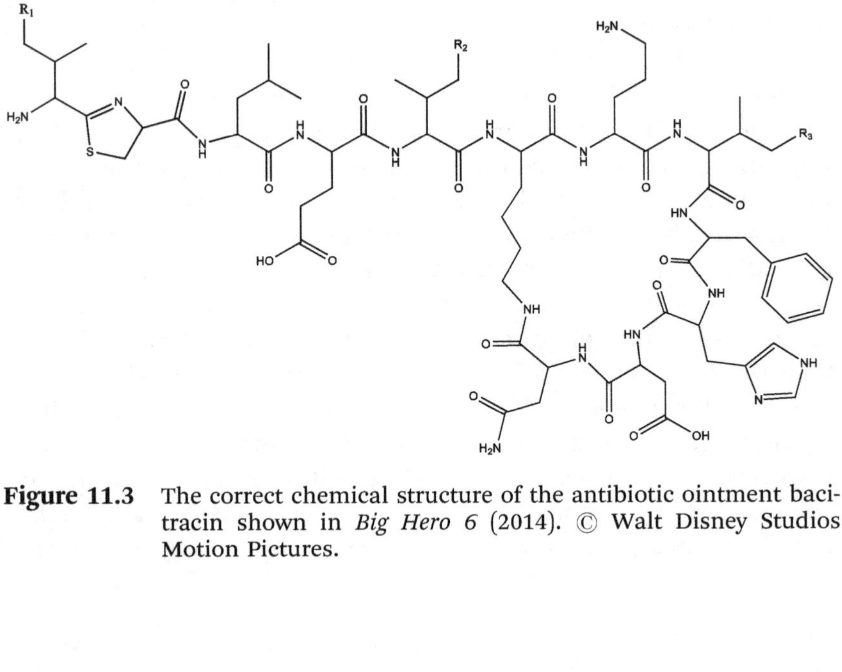

Figure 11.3   The correct chemical structure of the antibiotic ointment bacitracin shown in *Big Hero 6* (2014). © Walt Disney Studios Motion Pictures.

alongside some lab glassware. Doug enters the lab wearing a full-face breathing apparatus and a yellow hazmat outfit. He dumps the flowers into a metal reaction vessel and uses the elaborate glassware and tubing apparatus to concentrate the psychotropic drug. If the references to *Breaking Bad* weren't already obvious, he then mentions his sheep collaborators names as "Jesse" and "Woolter", a parody of the *Breaking Bad* protagonist Walter.

Next, we go in search of some *Soul* (2020), with a music teacher who falls into a coma following an accident and needs to reunite his separated soul with his body. While *en route* to "The Great Beyond", he ends up in "The Great Before" where he is mistaken as a mentor and paired with "soul number 22" who refuses to go to Earth. As they search through the "hall of everything" for a "spark", there is a brief scene in a chemistry classroom where soul 22 uses some test-tubes to create a small explosion. On the chalkboard behind them is the correct chemical structure and formula for caffeine ($C_7H_8N_4O_2$). Notably, both characters are wearing safety goggles for this scene.

So far, we have seen an improvement in the depiction of chemistry for younger audiences on the big screen compared to previous discussions. But what about the small screen? *Atom Town* (2022) is a community of colourful characters who represent the chemical elements of the periodic table. The core characters are based on the elements of all living things, namely Carbon, Hydrogen, Oxygen, and Nitrogen, with frequent appearances of Sulfur and Phosphorus as well as Calcium and Potassium. The personality and image of each character reflects the properties of the element they are named after, so Oxygen is bubbly, Carbon looks like coal, and Calcium is a milk carton (Figure 11.4).

Over time we meet several other elements who are all neighbours in an apartment complex which is shaped like the modern periodic table. The stories for each episode also contain references to real chemistry like potassium excelling at the banana eating competition, nitrogen tending to his crops, sodium running a restaurant with chlorine, and titanium showing off his immense strength. There is also an episode where the atomic weights of the elements are matched up on both sides of a see-saw. The elements also compete in a sports game called

Figure 11.4   The Carbon, Hydrogen and Oxygen characters from *Atom Town* (2023). Reproduced with permission from Turnip + Duck and Treehouse Republic.

"buckyball" and there's a heavy metal band with Lead, Mercury and others.

Finally, we return to the movies for *Elemental* (2023), based around the ancient properties of matter mentioned earlier in relation to the *Fifth Element* (1997), *i.e.* Earth, Wind, Fire and Water. The plot follows a Fire family who emigrate to Element City and face xenophobia from the Earth, Wind and Water people. Their daughter Ember Lumen strikes up a relationship with a Water person named Wade Ripple. The chemistry references are subtle but frequent, and the story evolves from scientific principles into romantic chemistry.

Ember frequently uses her talents to solve problems, like turning sand bags into glass to fix a water leak. In another scene, Ember creates a hot-air balloon using her flame so they can get a better view of the city, much to Wade's amazement. They also meet for a date at the Alkali Theatre and Ember's flame changes colour when she stands on different minerals, referencing the flame tests for metal ions commonly used in education. Wade also has scientific talents, using his body as a magnifying glass to create a flame from Ember's light. Finally, Wade evaporates when it gets too hot, but thankfully he condenses again later.

REFERENCES

1. A. Ju, *Life "not as we know it" possible on Saturn's moon Titan*, Cornell Chronicle, https://news.cornell.edu/stories/2015/02/life-not-we-know-it-possible-saturns-moon-titan.
2. A. Ulqinaku, G. Sarial-Abi, E. L. Kinsella and E. R. Igou, The Breaking Bad Effect: Priming with an Antihero Increases Sensation Seeking, *Br. J. Soc. Psychol.*, 2021, **60**(1), 294–315.
3. J. Hare, *Breaking Bad – poisoning gangsters with phosphine gas*, Education in Chemistry, https://edu.rsc.org/analysis/breaking-bad-poisoning-gangsters-with-phosphine-gas/3007373.article.
4. J. Hare, *Breaking Bad II – acid bath disposal of bodies*, Education in Chemistry, https://edu.rsc.org/analysis/breaking-bad-ii-acid-bath-disposal-of-bodies/3007374.article.
5. J. Hare, *Breaking Bad III – thermite break-in*, Education in Chemistry, https://edu.rsc.org/analysis/breaking-bad-iii-thermite-break-in/3007375.article.
6. J. Hare, *Breaking Bad IV – can a little crystal blow up a room?* Education in Chemistry, https://edu.rsc.org/analysis/breaking-bad-iv-can-a-little-crystal-blow-up-a-room/3007376.article.
7. H. W. Moser, G. V. Raymond, S. E. Lu, L. R. Muenz, A. B. Moser, J. Xu, R. O. Jones, D. J. Loes, E. R. Melhem, P. Dubey, L. Bezman, N. H. Brereton and A. Odone, Follow-up of 89 Asymptomatic Patients with Adrenoleukodystrophy Treated with Lorenzo's Oil, *Arch. Neurol.*, 2005, **62**(7), 1073–1080.
8. J. G. Goll and B. J. Woods, Teaching Chemistry Using the Movie Apollo 13, *J. Chem. Educ.*, 1999, **76**(2–4), 506–508.
9. J. Hare, *Apollo 13 – lithium hydroxide saves the day*, Education in Chemistry, https://edu.rsc.org/feature/apollo-13-lithium-hydroxide-saves-the-day-/3007380.article.
10. R. S. Hoffman, Thallium Toxicity and the Role of Prussian Blue in Therapy, *Toxicol. Rev.*, 2003, **22**(1), 29–40.

Problem Solving

12.1 CHEMISTRY HAS ALL THE SOLUTIONS

Previously, we've seen several examples of chemistry used for problem solving, which have slowly moved us away from accidents, side effects, dangerous chemicals, and recreational drugs. However, to date most of the problem-solving with chemistry has been depicted in true stories. We previously saw how chemistry has been the solution to a variety of real problems like health issues, building rockets, and saving astronauts. A knowledge of chemistry was beneficial and applying that knowledge led to solutions. Here we will see this theme evolve further into fictional stories, making its way into the unlikeliest of places.

One of those unlikely places is in *Legally Blonde* (2001). A law student, Elle (Reese Witherspoon), wins her first court case using her obscure chemistry knowledge of hair perms when she catches the witness lying about washing her hair. She proclaims "isn't it the first cardinal rule of perm maintenance that you are forbidden to wet your hair for at least 24 hours after getting a perm at the risk of deactivating the ammonium thioglycolate". This statement is entirely true and ammonium thioglycolate is also known as 'perm salt'.

However, there is also a somewhat failed example of problem solving in the episode *Flowers for Charlie* (2013) from *It's Always*

Onscreen Chemistry: The Portrayal of Chemical Science in Film and TV
By John O'Donoghue
© John O'Donoghue 2025
Published by the Royal Society of Chemistry, www.rsc.org

Sunny in Philadelphia (2005–present), featuring scientists who make Charlie smarter with a pill. Some colourful liquids in glassware and a very incorrect and bizarre chemical structure can be seen in the background (Figure 12.1). Later, Dee gets stuck to the glue on a rat trap which prompts Charlie to recommend "gasoline" to dissolve the glue. Mac agrees by saying "gasoline is a solvent, that's basic chemistry. We need to be careful though, gasoline is volatile". However, instead of using it as a solvent, they get distracted by inhaling the fumes. Gasoline (petrol) is an organic solvent and might dissolve the glue, but there are other solvents that would be better like acetone in nail varnish remover.

Next, we return to the ultimate problem solver in the rebooted series of *MacGyver* (2016–2021). Interestingly, this version of the spy series borrows some chemistry ideas from other TV series discussed earlier like 'Elephant Toothpaste' from *The Big Bang Theory* (2007–2019) in the episode *Pliers* (2016) and thermite from *Breaking Bad* (2008–2013) in *Father + Bride + Betrayal* (2019). However, like the original series, MacGyver occasionally constructs devices for testing chemicals in order to detect an explosive or a poison.

In the episode *Mac + Jack* (2018) he uses red cabbage to create a pH indicator to detect the presence of explosives. This is a well-known real method of making a pH indicator, often used by teachers in lessons about acids and bases. There is also an

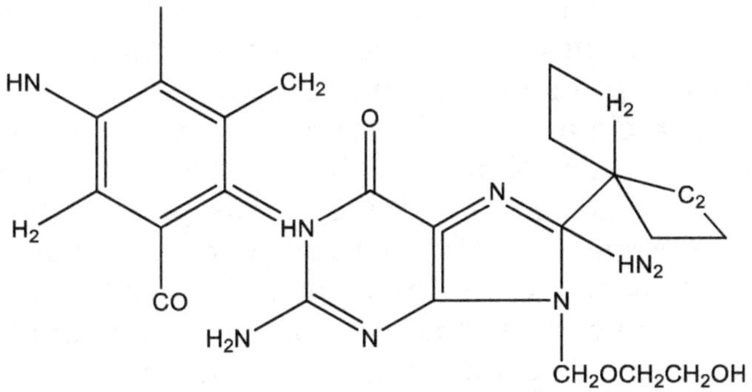

Figure 12.1 The impossible chemical structure as seen in *Flowers for Charlie* (2013) from *It's Always Sunny in Philadelphia* (2005–present). © 3 Arts Entertainment, RCG Productions, 20th Century Fox Television, FXP.

example of a homemade CO_2 tracker in *CO2 Sensor + Tree Branch* (2018), and he uses a fluorescent detergent and a UV light to detect cyanide poisoning in *Father + Bride + Betrayal* (2019). However, much like *Breaking Bad* (2008–2013) and the original *MacGyver* series, the level of chemistry is eventually toned down in later seasons. This may be due to fears that the public could repeat some of the unsavoury chemical creations, as previously mentioned in relation to *Fight Club* (1999).

Finally, we also have a new perspective on the 'chemistry causes explosions' trope when it is used to create a distraction in *Red Notice* (2021). In the laundry room of the prison, the protagonists explain "you know what happens when you mix soap, which is pure glycerin, with a cleaning solution that's essentially nitric acid? Nitroglycerin". They dump the mix into a tumble dryer which creates a large explosion, allowing them to escape. Although the throwback to *Fight Club* (1999) and other examples is interesting, the chemistry is greatly simplified. They would also need sulfuric acid and I'm not sure why a prison would be using nitric acid as a cleaner.

12.2   FORENSICS AND ANALYSIS

So far, we've only had sporadic examples of forensic chemistry on screen, but the turn of the millennium provides several new depictions starting with a brief mention of *The Bone Collector* (1999). The key onscreen chemistry scene is when the forensic scientist reports that the dirt on a bone they found is nitrogen rich, which indicates the presence of manure. However, because of the level of oxidation to nitrate, this indicates old manure, which leads them to the next clue. Away from the big screen we have the *CSI: Crime Scene Investigation* (2000–2015) and *NCIS: Naval Criminal Investigative Service* (2003–present) franchises. Both feature regular scenes in labs where the scientists perform analytical chemistry tests or "trace analysis", in addition to DNA, fingerprint, and ballistic analysis. Luminol also features, which is used to detect trace amounts of blood just like in real life.

The labs are usually shown with glassware filled with colourful and sometimes bubbling liquids. But they also include real tools like micropipettes as well as real instruments like liquid and gas chromatographs. However, these are often linked with exaggerated or misinterpreted techniques such as "electrostatic imaging". Just like *Medicine Man* (1992), the instruments usually provide instant answers to unknown substances with no baseline or standards used. Although real instruments can use large databases of spectra to identify known substances, unknown chemicals need several analytical techniques to work out a molecular structure.

On top of this, many of the chemical structures shown onscreen in both franchises are erroneous, like the one seen in the *NCIS* episode *Worst Nightmare* (2010) about a gas attack at a school. The gas is identified as "diethylmononitratetrisilicate potassium hydroxide" and the structure shown is a very large, disconnected, multi-ringed chemical structure featuring several errors. The head chemist explains that there are "very few chemical compounds that have the ability to knock out that many people" but admits that she has never heard of this gas. This would not be much of a surprise since it cannot exist based on the structure provided.

The novel nature of the compound leads the team to look for a chemist, but it is not explained how they managed to determine the chemical structure of such a complicated molecule so quickly. It's also worth noting that the use of erroneous chemical structures has not improved over the many seasons of *CSI* and *NCIS*, with a recent example (Figure 12.2) featuring a carbon attached to 14 hydrogens and an aromatic ring (the maximum number of bonds for carbon is 4) in the *NCIS: Sydney* episode *Extraction* (2024). These errors are in addition to the mispronunciation of chemical names across both franchises such as butan-2-one, with the 'one' pronounced like the number instead of like 'own' as it should be.

There is also a reference to *Breaking Bad* in the *CSI* episode *The Book of Shadows* (2014), about a high-school chemistry

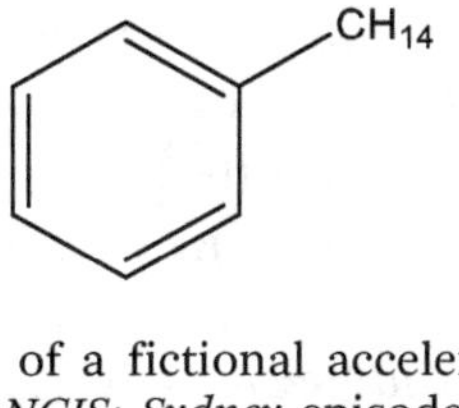

Figure 12.2 The structure of a fictional accelerant called "triptahydro" as shown in the *NCIS: Sydney* episode *Extraction* (2024), featuring an impossible carbon attached to 14 hydrogens and an aromatic ring. © Paramount+, Network 10.

teacher who is poisoned with cyanide. His death initially looks like a science experiment gone wrong, involving an explosion of potassium chlorate and sugar as previously seen in *October Sky* (1999). However, the equations on the board are a mixture of correct formulae like sucrose ($C_{12}H_{22}O_{11}$) coupled with a nonsensical equation: $4NaCl + 3H_2O \rightarrow H_6O_3 + Na_4 + Cl_4$. It is eventually revealed that the teacher was part of a Wiccan coven in a throwback to the 'evil biology teacher' from *The X-Files*. The forensics team also find red phosphorus at his home, but it turns out that he was using this for making spells, not meth. Finally, it transpires that a student is responsible for the poisoning and the teacher was producing steroids for the school athletes.

Following the success of *CSI* and *NCIS*, we also have *Bones* (2005–2017), about a forensic anthropologist (Emily Deschanel) and her team of scientists at the "Smithsonian Institute". Although most of the science is biological in nature, there are several episodes with chemistry such as *The Doom and Gloom* (2013), where phosphorus is found under the nail of a victim. Aluminium and exhaust fumes are also found on the clothing of victims in *Aliens in a Spaceship* (2006). The lab in *Bones* (2005–2017) often includes futuristic equipment such as a "holographic projector". The use of these futuristic devices gives the science a sci-fi feel, which also tells us that we need to suspend our disbelief about the universe in which it is set. This helps to overlook the various goofs like lab instruments that require little or no preparation of samples, and the determination of unknowns with little or no input from the scientists. It also seems that the magnets in MRI machines do not attract ferrous objects as they do in real life, since Booth (David Boreanaz) stands near one with a gun.

Similar errors can also be seen in *Dexter* (2006–2013), about a forensic technician with a secret parallel life as a vigilante. In *The Angel of Death* (2011) he concludes from the analysis of glue that it consists of "a polyvinyl acetate polymer dissolved in toluene". However, the molecule shown on screen is not polyvinyl acetate (PVA) (Figure 12.3). It seems strange that a lot of effort went into creating this fictional molecule, which doesn't have any chemical errors, only to misname it PVA. Also, worth a brief mention here as well is the *Rizzoli & Isles* (2010–2016) episode *Tears of a*

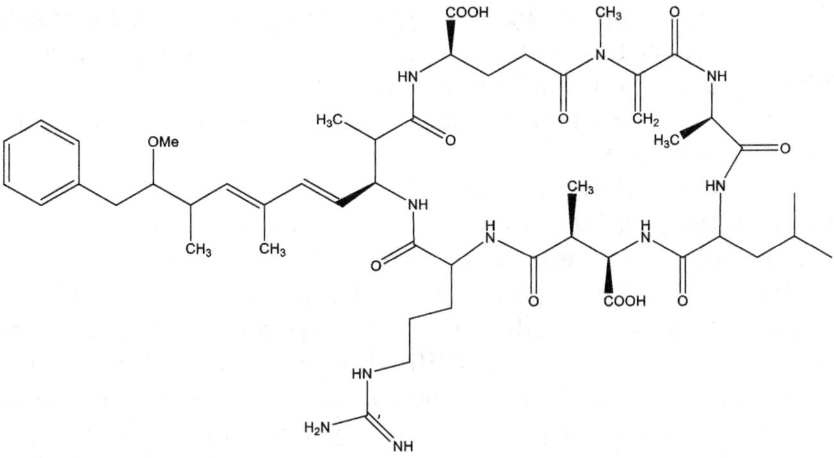

Figure 12.3   The elaborate and incorrect structure of PVA (polyvinyl acetate) as seen in the *Dexter* episode Angel of Death (2011). © 3 Arts Entertainment, RCG Productions, 20th Century Fox Television, FXP.

Clown (2014), where the team find "volatile chemicals" on a flower petal. A volatile substance will evaporate entirely, so there wouldn't be any left to find.

Overall, the three most popular chemical themes for onscreen forensics in the early 2000s are drugs, poisons, and explosives (accelerants). The most popular poison, other than cyanide, is thallium, previously discussed in *The Young Poisoner's Handbook* (1995). Thallium and its compounds are extremely toxic in real life, even contact with skin is dangerous since thallium(ɪ) compounds have a high aqueous solubility and are readily absorbed. It appears in the *NCIS* episode *Dead Man Walking* (2007) as thallium-laced cigars to murder a naval officer and again for a similar purpose in the *NCIS: Hawaii* episode *Spies* (2022). Radioactive thallium poisoning appears in the *CSI: NY* episode *Page Turner* (2008), and a thallium compound is used to poison a medical examiner in the *CSI: Vegas* episode *We All Fall Down* (2023).

Away from forensics, thallium also appears in medical series like in the *House* episode *Whatever it Takes* (2007), where it mimics polio and is later cured using a high dose of vitamin C. There are some real-world experiments showing vitamins C and E as helpful protective measures, but not as treatments.

Thallium also appears in the *Royal Pains* episode *Run, Hank Run* (2011) in the victims swimming pool water, which references the high aqueous solubility of thallium. The main symptoms are described onscreen as muscle spasms and nerve problems, which is accurate.[1]

Comedies and dramas have also featured thallium poisoning, such as *Drop Dead Diva* in the episode *Ashes to Ashes* (2012), when it is found in the cremation remains of a client's husband. The boiling point of thallium is 1473 °C, so finding it in the cremation remains is possible. Also, a biology professor researching how to prolong life has his life shortened by thallium in the *Elementary* episode *Nobody Lives Forever* (2018). The cause of death is determined by the dead lab rats near the professor's body, which ate some of his remains. It also transcends languages, appearing as a poison in the Hong Kong series *Two Steps from Heaven* (2016). On the big screen is also appears in *Big Nothing* (2006), *The Edge of Darkness* (2010) and *Spectre* (2015), which will be discussed later.

However, the most interesting depiction of thallium poisoning is in *Father Brown*, in the episode *The Two Deaths of Hercule Flambeau* (2018). It remains one of the few examples which mentions the real antidote, determined by Fr Brown through a series of clues. Prussian blue, or potassium ferric hexacyanoferrate, can be used to treat all forms of thallium poisoning in addition to radioactive caesium. It works by combining with the heavy metal in the intestines, forming a complex, which can then be removed in the stool.[1]

Sticking with the theme of mostly correct chemistry, next we explore some *Criminal Minds* (2005–present) in the episode *Sense Memory* (2011). The criminal here is a killer scientist who is obsessed with female scents. The victims are found with methanol in their lungs and in the next scene we get a close look at a distillation apparatus used for extractions. The camera starts with a large boiling flask which leads up to a condenser and into a separation funnel. When he opens the tap, it drips into a conical flask, which he then smells. Later, we see the entire apparatus with the glassware containing a cloudy colourless liquid.

The investigators describe methanol as the "simplest of the alcohols" and "ingesting 10 ml of it will cause permanent blindness, and as little as 30 is potentially fatal", which is accurate. A list of normal uses for methanol is also provided, all of which are correct. Eventually it transpires that the scientist is extracting the smells of his victims by boiling their bodies in methanol using salt and nitric acid. He also uses chloroform to knock out his victims, which gets him caught since chloroform is regulated, unlike methanol.

We then return to the big screen with *Sherlock Holmes* (2009), starring Robert Downey Jr as the famous detective. First, we have a paralysed victim who is unable to escape the boiling water in a copper bathtub. It is implied that the copper and water combination activated the paralytic agent, but no details are provided. Later, the villain fakes his death by ingesting a substance causing a temporary lack of pulse. Sherlock calls it "mad honey disease", which is true and originates from real-world grayanotoxin found in rhododendron plants. Finally, cyanide gas is part of the evil plan to poison the UK House of Lords, but the villain is immune due to something he drank earlier. In the real world, the vitamin B12 analogue hydroxocobalamin is effective as an antidote. Also, in the sequel *Sherlock Holmes and the Game of Shadows* (2011), he identifies the poison on a dart as curare using smell. We previously discussed this real poison in relation to the James Bond parody *Our Man Flint* (1966). However, it doesn't have any smell. It was named 'curare' from the Indian word for 'poison', and during the 19th century it was briefly used as a muscle relaxant to make surgery easier, which brought it to the attention of the Sherlock Holmes author Arthur Conan Doyle.

Around the same time, the famous detective also appears on the small screen in the British mystery *Sherlock Holmes* (2010–2017), starring Benedict Cumberbatch. There are frequent references to Sherlock's drug addiction and the famous 221B Baker Street flat contains a lab with glassware apparatus, usually containing colourful liquids. There are also several chemistry references throughout, starting with an unnamed poison in *A Study in Pink* (2010). Explosives also appear in several episodes, such as *The Empty Hearse* (2014), but we are never told what type of explosives they are. Sherlock also uses a medical lab in London's Barts Hospital to perform analysis, where he is sometimes shown using chemistry tools like pipettes and droppers.

There is also a reimagining of *The Hound of the Baskervilles* in *The Hounds of Baskerville* (2012) with 'H.O.U.N.D.' the code name for a hallucinogenic chemical weapon which turns people "insane". A whiteboard is briefly shown with several chemical formulae, but they do not contain subscripts. One is "C17H23NO3", which is atropine, the anticholinergic mentioned previously in *The Rock* (1996), and another is "C19H35NO2" which is dicycloverine, used to treat intestinal spasms. The formula "C17H21NO4" is also partially hidden under some sugar that Sherlock is testing, which corresponds to cocaine, the drug of choice used by Sherlock later in *The Abominable Bride* (2016). Later, it transpires that a chemist has continued to research the H.O.U.N.D. chemical weapon, due to his obsession with it. More recently, the *Enola Holmes* (2020) movie see Sherlock's teen sister put her hand to chemistry with her mother, using a glassware apparatus at home. In a brief flashback scene, they drop a colourless liquid onto a black powder which produces a flash, but the reaction isn't explained. Notably though, they are both wearing safety goggles.

However, it's not all bad: we have an excellent use of chemistry to solve a murder in the episode *Erupting Murder* (2017) from *Death in Paradise* (2011–present), about the death of a volcanologist. The inspector and his team find phenolphthalein in the victim's room, which he was using for "testing soil samples". However, the inspector incorrectly remembers that it "turns pink when an acid is added". First thought to be a heart attack, it turns out the victim died of hypoxia. Later, they find tanks labelled O_2 and CO_2 in the observatory and the inspector realises that he had it backwards – the natural colour of phenolphthalein is actually pink. Since the sample they found was colourless, this proves that the victim was killed by piping CO_2 into his room because it's an "acidic gas".

Finally, we finish this section with the miniseries *True Detective: Night Country* (2024), about a murder at an Arctic research station. One of the researchers is described as an "environmental chemist" and there are some scenes in a lab with glassware. The scientists are also referred to as "dorks" by one of the police officers. Later, we discover that the scientists are all responsible for a girl's death when she "destroys 2 years of work", meaning the scientists felt that their research was more important than a person's life. There is also a scene where the protagonist pours an unnamed solvent over a hatch and uses a UV light to reveal fingerprints. In real life, Sudan black is a dye which stains greasy finger marks.

12.3 CHEMISTRY IN SCI-FI

In March 2011 a magnitude 9.1 undersea earthquake occurred off the coast of Japan, resulting in a tsunami which resulted in the Fukushima Daiichi nuclear disaster. It was the most powerful earthquake ever recorded in Japan and the fourth most powerful in the world. The combination of electrical grid failure and damage to backup systems resulted in the release of radio-active contaminants into the environment, making it the worst nuclear accident since the Chernobyl disaster in 1986. Along with several other accidents in the early 2000s, it subsequently inspired a resurgence in onscreen radiation for fictional and true stories.

We start off with an example of a fictional workplace radiation exposure in the US dystopian thriller *Elysium* (2013). While working in an assembly plant for robots, Matt Damon's character gets trapped inside a malfunctioning radiation chamber. However, it is not explained what kind of radiation he is exposed to or why the robots are blasted with radiation during their assembly process. Perhaps it's some form of test for radiation resistance or maybe it's a curing process for paint. Regardless, this is also another example of 'evil industry' when the company provide him with non-descript pills and then leave him to die while the assembly line is resumed "as quick as possible".

In line with new exoskeleton inventions in the real world and following those seen in *Elysium* (2013), we then return to cyborg chemistry in a reboot of the classic *Robocop* (2014). Like the original movie, it is claimed that fully autonomous robots lack humanity compared to a robot built around a human. The robots "cannot feel, they cannot value human life". Also, like the original, this reboot explores the privatisation and automation of law enforcement with 'evil industry' attempting to pull the strings. However, it then transpires that a cyborg is inferior to the reaction capabilities of a full robot since a cyborg retains fear and emotion.

To fix "the human issue", they attempt to remove much of his autonomy by controlling his dopamine levels, only giving him the "illusion of free will". There are also references throughout to the Tin Man from *The Wizard of Oz* (1939), including the song *If I Only Had a Heart*. The scientists are nearly always seen in lab coats at the OmniCorp industrial labs. The theme of recreational drugs also makes a reappearance, but the drug labs are more like warehouses with crates of drugs. Notably though, when OmniCorp attempts to kill Robocop to "remove the problem", the scientists turn against the company to help him. Also, the media are always shown on the side of the company and against the scientists.

Sticking with the theme of cyborgs, a few years later we have another entry in the *Terminator* franchise with *Terminator: Dark Fate* (2019). This time the protagonist is rescued by a "cybernetically enhanced human", called an "Augment". The enhancements result in an accelerated metabolism, requiring the Augment (Mackenzie Davis) to take medical injections when overheated after an intense battle. After the first fight sequence, there is a scene in a pharmacy with the Augment asking for "any anti-convulsive, sodium polystyrene sulfonate, insulin, benzodiazepine", all of which are real medicinal drugs. Sodium polystyrene sulfonate is used to treat high levels of potassium in the blood, known as hyperkalemia. Benzodiazepine is a type of sedative which slows down body and brain functions. The Augment is also powered by a "thorium microreactor", a real reactor that is currently being researched around the world.[2]

Next, we return to space chemistry with one of the best examples of problem-solving using chemistry in *The Martian* (2015). The protagonist (Matt Damon) has been abandoned on the planet Mars and needs water to grow crops. Starting with hydrazine (N_2H_4), a real rocket fuel, he drips it over an iridium catalyst which correctly generates hydrogen (H_2) and nitrogen (N_2) gas. Then, by burning the hydrogen in the presence of oxygen, it creates water vapour. Additionally, burning the hydrogen prevents the formation of ammonia which is produced as a side product from the decomposition of hydrazine.

The chemistry is entirely correct and commendable, but it isn't explained where he gets the iridium catalyst from since it is one of the rarest elements in the Earth's crust. Initially he miscalculates the amount of oxygen which results in a typical 'chemistry causes explosions' trope as he blows himself up, but it does add some humour to the scene. Also, the toxicity of hydrazine is not mentioned – he would need a mask or a hazmat suit to work with it. Despite these caveats, this scene contrasts heavily with the brief reference in *Apollo 13* (1995), making this a good example of how onscreen chemistry has evolved over time.

Next, we return to a dystopian future in *Blade Runner 2049* (2017), featuring spectral analysis used to identify traces of tritium. Tritium is a real but rare radioactive isotope of hydrogen with a half-life of about 12.3 years. It is produced naturally in the upper atmosphere when cosmic rays strike nitrogen molecules. However, relevant to the movie, it is also produced during nuclear explosions, and as a byproduct in nuclear reactors. It is explained in the movie that it was due to a dirty bomb which exploded outside of the city of Las Vegas.

Finally, we finish this discussion with a brief scene of what appears to be an illegal drug lab in the second part of the *Star Wars* miniseries *Obi-Wan Kenobi* (2022). The lab is filled with glassware containing bubbling colourful liquids and the workers are wearing colourful overalls, gloves, masks, and safety goggles. There is also a reference to *The Man in the White Suit* (1951) through a brief use of the same melodic bubbling sound. The protagonist uses a powder to cause one of the boiling flasks to explode, creating a distraction so he can access the room beyond.

12.4 EXPOSING THE TRUTH

Previously we discussed numerous true stories, biographies, and fictionalisations of true events, many of which showcased the use of chemistry for problem solving. In recent years there has been a resurgence in true stories featuring onscreen chemistry, however they still revolve around 'side effects' and 'evil industry' or 'greed'. We start here with a return to radiation in a real workplace with a true story about 'evil industry' in the US drama *Radium Girls* (2018), featuring several similarities to *Silkwood* (1983).

The plot follows female factory workers who contract radiation poisoning from hand-painting radium dials on clock faces during the 1920s. This was done to make the dials glow in the dark. The movie opens with a travelling salesman flogging "radioactive water, Radithor" which was a real quack pseudo-scientific product from this time. Radium is described as a miracle of the twentieth century which was "discovered by Marie Curie", and there is brief image of the correct electron configuration for radium. The factory workers are also instructed to point their brushes on their lips to give the brush a fine tip for accurately painting the dials.

Later, one of the girls becomes sick and loses a tooth but the company doctor tells her that she is healthy and that her aliments are due to "poor hygiene". After consulting with an activist, they learn that radium "causes tissue damage" and when the victims die, the company doctor claims that they succumbed to syphilis. The embarrassment of contracting the sexually transmitted disease means the women don't talk about it. Later, the two protagonists allow the activists to exhume their sister's body and discover that "the levels of radioactivity in Mary's bones are 1000 times the safety limit". When these details are brought to the company, they say that the paint is harmless. After a lengthy case of denial from American Radium, the workers eventually won damages in 1938.[5]

Sticking with the theme of radiation, next we return to the famous chemist Marie Curie (Rosamund Pike) in the British biopic *Radioactive* (2019). Initially, the other scientists refuse to share a lab with her because her equipment is taking up too much space. We then have a strange example of mocking teachers when she proclaims "science is lost to me, I'm going to have to become a teacher", to which her sister proclaims "you are impossibly dramatic". It is unlikely that the real Marie would have said this, since she was the daughter of two teachers and later became a lecturer herself. While looking for a new scientific home she meets Pierre who is watching a girl dancing with a colour changing dress. This is a reference to *Annabelle's Serpentine Dance* (1895), as discussed previously, which accurately matches when they met and got married in real life. He tells her that she is one of only twenty-three female scientists in the university science department, to which she replies "a prime number".

Later, we see her conducting an experiment with glassware in her apartment which almost results in a fire, convincing her to join Pierre's lab. His lab is shown with glassware, wooden benches, and cabinets. Some of the volumetric flasks in the background also contain colourful liquids, as expected. Later, she performs a hot filtration, *i.e.* pouring a hot substance from a conical flask into a funnel with some filter paper. After several experiments she begins to suspect that an undiscovered element is mixed with her uranium samples, so Pierre gifts her a "quadrant electrometer, capable of measuring precisely the tiny amounts of electric charge". Initially, Marie resists Pierre's suggestions for a professional partnership by insisting on working alone. But it is their partnership that eventually leads to the discovery of two chemical elements (polonium and radium).

There is also a scene explaining how they will extract the elements from the pitchblende, which includes crushing and boiling in addition to acid and alkaline extractions. However, the chemistry mostly ends there, with significantly less detail provided compared to *Madame Curie* (1943). The immense work of extracting the tiniest amount of radium from four tonnes of pitchblende is only a short montage. However, we do see a green glow this time, which is an update from the previous glow seen in black and white. The rest of the movie mostly focuses on her legacy, which includes the use of radium in products like toothpaste and face powder, as well as atomic bombs and the Chernobyl nuclear meltdown. This is contrasted with the use of radium chloride to shrink a tumour and the first radiation therapy for retinoblastoma in 1957.

However, it is not true that Pierre travelled alone to collect the Nobel Prize in Physics in 1903 – neither of the Curies travelled in real life. It was collected by their collaborator Henri Becquerel, who is only briefly mentioned in this version. In *Radioactive* (2019) this causes an argument, and it is claimed that she didn't travel because she just gave birth, which is also not true. Her daughters were born in 1897 and 1904, and the Curies travelled together to collect the Nobel Prize in 1905. However, unlike the previous movie, this version includes her winning the Nobel Prize in Chemistry in 1911 and her mobile X-ray vehicles on the battlefields of WW1. There is also an accurate periodic table from this era behind her as she teaches at the University of Paris.

Once again sticking with the theme of radiation, in the same year we also have the miniseries *Chernobyl* (2019), which charts the true story of the real-world nuclear meltdown. In the second part of the series, *Please Remain Calm* (2019), there is an accurate periodic table in the background from that time. Later, there is also a clear explanation of how a nuclear reactor works which includes the fuel rods, control rods, and water. He also goes on to explain the issue, *i.e.* the element xenon has built up in the reactor, poisoning the control rods.

We then return to 'evil industry' with a story that shares numerous similarities with *Erin Brockovich* (2000). *Dark Waters* (2019) tells the story of one of the largest legal settlements in US history, over the dumping of byproducts from the manufacturing of non-stick polytetrafluoroethylene (PTFE). The protagonist Robert Bilott (Mark Ruffalo) is a lawyer who takes on a case from his homeland. After meeting a farmer about his dead cattle, he meets a scientist to get more information about PFOA (perfluorooctanoic acid), but the scientist only knows about PFOS (perfluorooctanesulfonic acid), both of which are abbreviated to "C8". Although *Dark Waters* (2019) also uses a 'parachute scientist', the chemistry depicted is correct and significantly more detailed than the same scene in *Erin Brockovich* (2000).

The parachute scientist correctly describes the substances as "long chain fluorocarbon, synthetic", to which Bilott responds "I'm sorry, but chemistry was my worst class in high school". This prompts the scientist to further explain the concept with "synthetic, man-made, Frankenstein, and long-chain fluorocarbon is a sequence of carbon atoms with a fluoride". He then draws a chain of carbon atoms in his notebook, eight in total, corresponding to C8. He goes on to say "a chain like that is pretty much unbreakable, biochemically speaking". In the real world it was dumped into rivers for many decades. Bilott's work triggered an epidemiological study which found a probable link between PFOA/PFOS and several health conditions. It is estimated that PFOA and PFOS are in nearly everyone's bloodstream, but it has decreased since the global ban in 2013.

Sticking with the theme of 'evil industry', we also have two dramatizations about the potent painkiller medication oxycodone in *Dopesick* (2021) and *Painkiller* (2023). Both tell the story of how a profit-driven pharmaceutical company contributed to the opioid epidemic in the US through the marketing and overprescription of oxycodone. Both series are based on the books of the same names and represent fictionalisations of the true story. However, both take a different approach at humanising the tragedy, with examples of different characters taking the painkiller due to work-related injuries. Like other opioids (*e.g.* heroin, morphine) oxycodone is addictive, but it was marketed to doctors as "not like the others". However, neither series contain much in the way of onscreen chemistry, unfortunately, other than surface-level references.

Next, we have the fictionalised true story of the convicted fraudster Elizabeth Holmes (Amanda Seyfried) and her disgraced biotechnology company Theranos in *The Dropout* (2023). In the miniseries, she claims that her revolutionary device can perform blood tests without the need for medical labs. The screen adaption of her story also stars Stephen Fry as the pioneering real-life British biochemist Ian Gibbons. In keeping with the style of lab glassware apparatus now commonly seen onscreen, several scenes feature an elaborate setup of conical flasks, boiling flasks, condensers, and other items joined together with lots of nonsensical tubing. The entire apparatus is also filled with green liquid.

In the third part of the miniseries, the whiteboard behind Elizabeth contains a random array of molecules and diagrams. Most look chemically correct, but there are some mistakes like an impossible triple- and quadruple-bonded oxygen. However, *The Dropout* (2023) is one of the few depictions where incorrect onscreen chemistry is somewhat acceptable because the story is about fraudulent science, so fake chemistry is what they were peddling. Stephen Fry's character refers to himself as the "head of chemistry" and all the scientists believe they are doing real work. While trying to sell the concept to a large retailer, Elizabeth attempts to hide all the lab staff because she doesn't trust them. The scientists are mostly presented as honest. Because of their honesty and her mistrust, she makes them sign strong NDAs (non-disclosure agreements) to prevent them from ever talking about their work.

In the same year we also have *The Railway Men* (2023), about the tragic Bhopal chemical accident of 1984. Previously there was a movie about the same topic called *Bhopal Express* (2001), but it contained no mention of chemistry, only the word "gas". In the miniseries, several examples of lax safety standards are shown, as well as senior staff downplaying concerns raised by technical staff at the Union Carbide plant. They fire the staff member who hit the alarm because of a faulty gauge which indicated high pressure in the tanks labelled "reflux". Later, it is accurately explained that the plant makes pesticides from methyl iso-cyanate (MIC), described as "liquid dynamite, the most danger-ous and volatile chemicals in the factory". It is also mentioned that MIC reacts violently with water and gives off large quantities of heat. The effects of MIC poisoning are correctly portrayed as foaming at the mouth and coughing up blood. An autopsy also shows discolouration of the lungs, coupled with a smell of al-monds which indicates the formation of hydrogen cyanide.

Finally for this section, we finish here with the biography *Oppenheimer* (2023), which gives us another look at the life of the titular physicist. During a flashback, there is a brief scene in a chemistry lab where we see a bottle labelled "potassium cyan-ide". His chemistry lab work leaves little to be desired, so his tutor forces him to clean up the lab while everyone else attends a lecture by the famed Danish physicist Niels Bohr. In a reference to *Snow White*, he uses a syringe to inject the poison into an apple for his tutor as revenge. However, the next day Niels Bohr nearly eats it but Oppenheimer knocks it out of his hand at the last minute. It is said that potassium cyanide has a bitter taste, so perhaps the sweet taste of an apple might hide that. The 'cyanide apple' is also sadly associated with the true story of the British computer scientist Alan Turing, whose story was told in *The Imitation Game* (2014).

12.5   BREAKING BONDS

We finish our discussion on 'Bonding' with Daniel Craig's gritty reboot of the immortal spy in *Casino Royale* (2006). After some parkour and hand-to-hand combat, Bond is later poisoned at a casino when his favourite martini is spiked. As he begins to feel unwell, he stumbles out to his car where he attaches a defibril-lator to his chest, which also allows him to communicate with HQ. Due to the ventricular tachycardia that he is experiencing, the team conclude that he is a victim of digitalis poisoning. Digitalis does cause ventricular tachycardia, which is a type of abnormal heart rhythm resulting in the body not receiving en-ough oxygenated blood. However, like in many movies, the timing is sped up for dramatic effect. Poisoning from digitalis takes hours to show is effects, whereas the movie implies that the effects are almost immediate. During the tense scene, Bond also injects himself with lidocaine using a "blue combi pen", which would have sufficed without the use of a defibrillator.[3] Nonetheless, this is one of the most accurate examples of onscreen chemistry in the franchise.[4]

Next, we return to the theme of energy in *Quantum of Solace* (2008), with some throwbacks to *The Man with the Golden Gun* (1974) discussed earlier. This time the villain (Dominic Greene) appears to be staging a *coup d'état* in Bolivia to access the country's natural oil reserves. However, the real plan involves restricting the country's fresh-water supply to create a monopoly. Later we learn that his remote desert hotel is powered by "hydrogen fuel cells", which are described as "unstable". However, the link to the water hoarding is never made explicit, although presumably he needs the water to produce hydrogen *via* electrolysis. There are several references to the West's reliance on oil with the hotel representing energy independence, but apparently this comes with its own costs. The tanks labelled "hydrogen" provide an explosive finale and reinforce the earlier comment about instability.

We then parachute into *Skyfall* (2012) with some exaggerated computer hacking. The key chemistry scene is when the villain (Javier Bardem), an ex-British spy, explains that he used the "hydrogen cyanide capsule" embedded in his tooth during torture. But instead of ending his life, it "melted" his jaw and left him horribly disfigured. Hydrogen cyanide (HCN) is a fast-acting poison with a boiling point of 25 °C, considerably less than body temperature of 37 °C. This makes it a gas normally, meaning it would be difficult to store in a tooth. But even if that was possible, the amount stored in a tooth would not be enough to kill or cause disfigurement. This also represents another example of misunderstanding corrosion, since HCN is actually a weak acid. Normally 'suicide pills' (also called L-pills) contain concentrated potassium cyanide (KCN), not HCN, which absorbs internally and forms HCN in stomach acid.

Also worth a brief mention is the use of depleted uranium shells, which are used as armour-piercing bullets in real life because they are 68% denser than lead. Although they do contain tiny amounts of the highly radioactive isotope U^{235}, they mostly consist of the less radioactive U^{238}, hence the term 'depleted'. In the next outing, *Spectre* (2015), Daniel Criag broke his

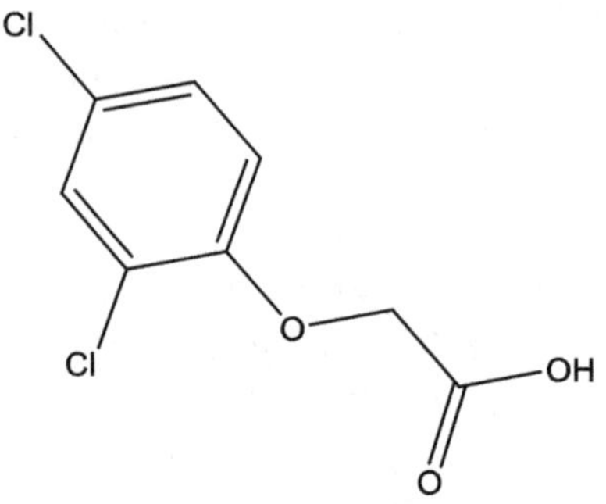

Figure 12.4 The correct structure of 2,4-dichlorophenoxyacetic acid, a synthetic herbicide which is one of two components of 'Agent Orange', as seen in *No Time to Die* (2021). © Eon Productions, Metro-Goldwyn-Mayer, Universal Pictures.

leg in real life but continued filming, even his stunt scenes. As mentioned previously, one of the people Bond meets is dying of thallium poisoning, administered to him by an assassin from the secret organisation Spectre. However, before the worst effects of the poison overcome him, he commits suicide. Also worth a brief mention is when Q analyses the Spectre ring, the computer reports that it is made of silver with the chemical symbol Ag appearing briefly.

Finally, we finish our discussion about Bonds with *No Time to Die* (2021). In a throwback to Craig's first outing in *Casino Royale* (2006), the villain (Rami Malek) uses a modified and aerosolised version of digitalis to selectively poison people *via* DNA matching. The villain's backstory also reveals that he was poisoned with a dioxin poison extracted from Agent Orange (Figure 12.4), disfiguring his face similar to the real-life 2004 poisoning of Viktor Yushchenko. Dioxin also causes reduced mobility and slow movement in real life, which Malek portrays throughout. In the end, we discover that all Bonds can be broken eventually, since 007 may not be immortal after all.

12.6 DRAMA AND ROMANCE

We start the final section here with a brief return to black-and-white lab scenes in the Spanish psychological drama *Talk to Her* (2002). The plot revolves around a comatose patient and a male nurse who abuses her. The lab scene is a 'film within a film' throwback to early silent horror movies like *Dr Jekyll and Mr Hyde*. A male scientist and a female scientist are surrounded by a very elaborate glassware apparatus with tubing and there are also several conical flasks filled with bubbling liquids producing smoke/condensate. The female scientist opens a tap on a separation funnel to fill a flask, which she then pours into a funnel. Eventually this produces a dark liquid which the male scientist takes from her and drinks. However, before he drinks it, she warns him "it could be dangerous, I haven't tested it on human beings yet". The formula shrinks him, but she cannot find an antidote.[†]

Next, is a brief discussion about *The Life of David Gale* (2003), which revolves around a former college professor on death row. The plot follows a journalist who explores the events leading up to his death sentence, which includes his activism against the death penalty. She eventually discovers a video recording that will prove his innocence, but she is too late. The lethal injection is described as a mixture of pancuronium bromide, which collapses the diaphragm, and potassium chloride, which stops the heart when given as an overdose.

Next, we skip forward to *Duplicity* (2009), which follows two corporate spies with a romantic history. When they first meet, Claire (Julia Roberts) drugs Ray (Clive Owen) so she can steal some documents, but it is not explained what drug she uses. Later, they become spies for competing pharmaceutical companies, one of which has developed a "cure for baldness". The chemical formula is stolen by the other company and both protagonists try to sell it to a separate Swiss company. However, it turns out that the original inventor of the formula has conned them and the formula they possess is just a "common lotion". The chemical structure shown on screen (Figure 12.5) is real and similar to common substances used in cosmetics and lubricants. Cyclopentasiloxane improves texture with a smooth and silky feel. Interestingly, a knowledge of chemistry would have saved both of them the embarrassment of being told that the formula is worthless.

[†] A word of warning, *Talk to Her* (2002) contains references to sexual abuse and scenes of nudity.

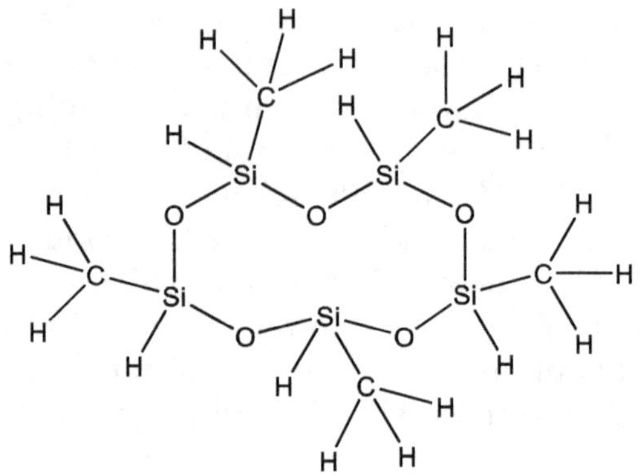

Figure 12.5 The claimed "cure for baldness", later called a "common lotion" reproduced from *Duplicity* (2009). © Relativity Pictures, Universal Pictures.

We then move onto *Better Living Through Chemistry* (2014), a comedy romantic drama about a pharmacist who starts an affair and 'breaks bad' by abusing prescription drugs. The pharmacy is shown with plenty of chemical containers and pills. There are also several mentions of over-the-counter and prescription medications like Lipitor, amoxicillin, Zoloft, Xanax and others. Eventually he starts mixing (splicing) various medications together, shown onscreen as a mortar and pestle crushing tablets. He also comes up with a formulation of his own creation to win the annual bike race. Later, he suggests to his mistress that he could alter her husband's heart medication to get around the prenup.

The title of the movie is a variant of the original slogan used by the chemical company DuPont who started using "Better Things for Better Living... Through Chemistry" in 1935. However, the phrase 'Better Living Through Chemistry' was later adopted by the counterculture movement in the 1960s and 1970s as a euphemistic reference to the recreational use of drugs. Relevant and interesting to the discussion of 'evil industry', DuPont claimed that their slogan was aimed at changing "opinions about the role of business in society", to address "unspoken fears of 'bigness' in business". DuPont eventually dropped the 'Through Chemistry' part of the slogan in 1982 and changed their slogan entirely in 1999 to "The Miracles of Science".[6]

Next, we have the US thriller drama *Old* (2021), based on the book *Sandcastle* by Pierre Oscar Lévy, about a secluded beach which rapidly ages people by a year every 30 minutes. It is later revealed that the beach is a front for a research team from an unethical pharmaceutical company. The team are conducting clinical trials of new medical drugs, which are administered to the guests by spiking their drinks. Because the beach accelerates the lives of the guests, the researchers can complete the lifelong drug trials within a day.

Worth a brief mention here also is Wes Anderson's comedy-drama *Asteroid City* (2023), set in a retro-futuristic version of the 1950s. The plot revolves around a documentary about the creation and production of a play. Most noteworthy is the focus on inventions, referencing 1950s movies, and a brief scene with a giant image of the "Periodic Table of The Elements". The chart is dated 1955 with nobelium (102) labelled 'Sz', presumably shorthand for 'synthesizing', which is correct. However, boron and aluminium are included in the s-block instead of the p-block, which is entirely incorrect, and the format of the table seems to be based on Remy's long-form version which wasn't published until 1956. The version normally used in classrooms and textbooks during this time was known as 'short-form' with the d-block elements integrated through doubling, not separated.

REFERENCES

1. R. S. Hoffman, Thallium Toxicity and the Role of Prussian Blue in Therapy, *Toxicol. Rev.*, 2003, **22**(1), 29–40.
2. U. E. Humphrey and M. U. Khandaker, Viability of Thorium-Based Nuclear Fuel Cycle for the next Generation Nuclear Reactor: Issues and Prospects, *Renewable Sustainable Energy Rev.*, 2018, **97**, 259–275.
3. T. Livanou, E. Voridis, K. Kaplanidis and C. J. Miras, Digitalis Toxicity, *Lancet*, 1972, **299**(7758), 1027.
4. K. Harkup, *The name's bond, chemical bond*, Chemistry World, https://www.chemistryworld.com/features/the-names-bond-chemical-bond/4012701.article.
5. K. Moore, *The Radium Girls: The Dark Story of America's Shining Women*, Sourcebooks, 2017.
6. A. McCarthy, *The Citizen Machine: Governing by Television in 1950s America*, 2010.

SEASON 13

Chiral Chemistry

13.1 THE TWO FACES OF CHEMISTRY

The cautionary tale of good and evil from the 1886 novel *The Strange Case of Dr Jekyll* and *Mr Hyde* has dominated the story of onscreen chemistry throughout this book. The Scottish author of the novel, Robert Louis Stevenson, also wrote other well-known and influential works including *Treasure Island* (1883) and *Kidnapped* (1886). Robert Lewis Balfour Stevenson was educated in Edinburgh, but later travelled extensively across France, California, and Polynesia. He eventually settled in Samoa, which became his final resting place at the age of only 44 after succumbing to long-term illnesses.

His father was a civil engineer and meteorologist, credited with 'Stevenson's formula' for calculating wave heights. However, he was named after his grandfather Robert, also a famous civil engineer who designed and oversaw the construction of lighthouses all over Scotland. His uncles were also civil engineers, in addition to his maternal grandfather, Thomas Smith, who was the first chief of the Scottish Northern Lighthouse Board. His mother Margaret Isabella Balfour was from a long line of gentry and ministers of the Church of Scotland. After his schooling he entered the University of Edinburgh to study engineering but showed little interest in the subject.

Onscreen Chemistry: The Portrayal of Chemical Science in Film and TV
By John O'Donoghue
© John O'Donoghue 2025
Published by the Royal Society of Chemistry, www.rsc.org

Instead, similar to the plot of *Dead Poets Society* (1989), he preoccupied his time with the university's Speculative Society. Here, members engaged in public speaking and literary composition as well as amateur drama productions. However, unlike the fictional movie, Robert's father encouraged his writing and paid for the printing of his first publication. In return, at the request of his father, Robert switched from engineering to law as 'security'. He later wrote a poem about turning from the family profession with the self-deprecating lines: "Say not of me that weakly I declined, The labours of my sires, and fled the sea, The towers we founded and the lamps we lit, To play at home with paper like a child". It should be noted, however, that Robert wasn't the only one to turn from the family profession. Although he was an only child, he had a close relationship with his cousin, who also turned from engineering to study art.

As well as rejecting an established profession, he also rejected Christianity and briefly declared himself an atheist. This was a huge disappointment to both his parents, especially his mother. In California he struggled financially and suffered ill health throughout, nearing death's door on several occasions. He later married an American magazine short-story writer and divorcee with children, Fanny Van de Grift. Despite his mother's disappointment at his religious views, she joined them in Samoa a few years before he passed away.

Returning to the topic of onscreen chemistry, the question is: did the story of Dr Jekyll and Mr Hyde originate from the world in which Stevenson lived? One of the proposed influences for the story is Stevenson's friendship with several homosexual men whose duality and oppression of their sexuality may have shaped the story.[1] In the novel, the protagonist is actually Jekyll's friend, a legal practitioner, which he no doubt based on his own legal background. The story unfolds as the protagonist investigates a series of occurrences between Dr Jekyll and a criminal known as Mr Hyde. For the first half of the book, both personalities are referred to as different people, with little indication that they are in fact the same person.

Jekyll's obsession with his lab work is only hinted at when the protagonist claims that Jekyll was secluding himself for many weeks. This prompts the protagonist to visit Jekyll, who only speaks to him through the laboratory window. This also happens to be one of only a few references to the lab. Also, the fact that Jekyll's seclusion arouses suspicion implies that it is out of character for him. As a result, there is no reason to believe that Jekyll was obsessed with his research, since this seclusion may be due to the time spent by him as Mr Hyde. The theme of obsession or perseverance with research therefore seems to originate from the movies, not from the original source material. Later, when he is unable to contact Jekyll, the protagonist breaks into the lab where he discovers Mr Hyde's body wearing Jekyll's clothes. This is the proof that the two characters are actually the same person.

A letter left by Jekyll reveals the truth with references to chemicals and the serum needed to transform in both directions, *i.e.* to induce the transformation from Jekyll to Hyde and back again from Hyde to Jekyll. It is explained that the supply of the salt used in the original serum ran low, and subsequent batches failed to work. Jekyll speculates that the original supply of the salt contained an unknown impurity which made the serum work. Eventually, he could not change back from Hyde and decided to commit suicide. Therefore, the cautionary warnings about conducting scientific research without considering all the consequences are only implied in the novel, not explicit.[2] As mentioned earlier, the Michael Caine version of *Jekyll* and *Hyde* (1990) is one of the few screen adaptions which sticks closely to the original story. But even then, it was still influenced by the earlier screen adaptions.

The earliest screen adaptations of Stevenson's story appeared at a time when chemistry was beginning to form two faces, one of good and one of evil, in the wake of the horrors of WW1. The duality of chemistry in the real world during the turn of the 20th century is perfectly encapsulated through the chemical element chlorine. Chlorine was first used to sterilise swimming pools around the same time that it was used to kill people on the fields of Flanders during WW1. Before the use of chlorine, pools and public baths were only cleaned through filtration and backwashing. Roman soldiers with wounds were advised not to use public baths for fear that their wounds could become gangrenous.[3]

The British scientist Sims Woodhead was the first to use chlorine to disinfect water, using a 'bleach solution' as a sterilising agent during the 1897 typhoid epidemic. This was later brought to the US in about 1908 and eventually made it into swimming pools in 1910, starting with the pool at Brown University. However, it should be noted that chlorine wasn't the only disinfectant used for swimming pools, ozone was also used in France and the Netherlands as early as 1893. But ozone required expensive equipment, whereas chlorine was in abundant supply as a byproduct from the manufacturing of caustic soda[4] (Figure 13.1).

The famous German chemist Fritz Haber won the 1918 Nobel Prize in Chemistry with Carl Bosch for their work on the production of ammonia as a fertilizer. However, around the same time, he also personally supervised the first use of chlorine gas

Figure 13.1   A "Swim for health in safe and pure pools" poster from the US, which was used for the Cleveland Division of Health to promote swimming as healthy exercise around 1940. © Science History Images/Alamy Stock Photo.

as a chemical weapon in 1915. He even supervised a gas attack on the day after his wife Clara died of a gunshot wound to the chest. The movie about Clara's life appears as the German-language biography *Clara Immerwahr* (2014) and will be discussed in more detail later.

Therefore, the conflict of 'good and evil' in the original story of Dr Jekyll and Mr Hyde seems to have been extrapolated to represent the two conflicting sides of chemistry. Looking back through all the examples in this book, on the one hand, chemists are mostly depicted as heroes and problem solvers with a desire to make people's lives better. There are of course some exceptions, like the chemist in *The Invisible Man* (1933), who wants fame and personal riches. Also, for the most part, the chemistry shown onscreen is usually depicted as beneficial to humankind, *e.g.* producing treatments for debilitating illnesses, green energy and making cellulose nitrate fireproof.

However, on the other hand, the benefits offered by chemistry also result in unexpected side effects like those seen in *Silkwood* (1983), *Erin Brockovich* (2000) and *Dark Waters* (2019), among many others. This is in addition to the purposeful use of chemistry to inflict harm through poisons like those used in *The Young Poisoner's Handbook* (1995) and chemical weapons as seen in *The Rock* (1996) and *Clara Immerwahr* (2014). The use of chemical weapons in WW1 is also depicted in the western romance *Legends of the Fall* (1994), when Samuel (Henry Thomas) is incapacitated by a yellow-green gas as his brother (Brad Pitt) attempts to save him. The gas does not immediately kill him, since the movement of the gas is accurately depicted as slowly dissipating around them. It does however blind Samuel, which means it is probably mustard gas rather than chlorine or phosgene in this case.

However, it should be noted here that before it appeared on screen in *The Young Poisoner's Handbook* (1995) and others, thallium first found its way into fiction through novels such as *The Pale Horse* by Agatha Christie in 1961. Agatha Christie had a background in chemistry, so real poisons tended to appear in much of her written work long before they appeared on screen. So, although screen adaptions of chemistry stories have added some artistic license and ideas of their own, written works have been hugely influential overall.

13.2 RISE, FALL AND RISE AGAIN

As discussed several times, most of the early depictions of onscreen of chemistry were linked with popular gothic horror stories like that of Dr Victor Frankenstein. But it is important to note that these stories were usually inspired by real-world scientific discoveries like that of Galvani. As a result, in a roundabout way, the first onscreen chemistry depictions were also linked with real-world discoveries. Then, during and after WW2, onscreen chemistry evolved into adventures and screwball comedies, which overall made chemistry appear fun. Chemists were also portrayed as heros and problem solvers like in *The Adventures of Tartu* (1943), and the onscreen chemistry was creative as seen in *It Happens Every Spring* (1949), *The Man in the White Suit* (1951) and *Monkey Business* (1952). However, despite all the positivity onscreen, in the real world, the evil side of chemistry was still lurking in the form of recreational drugs and industrial accidents.

Although opium was long known by the Sumerians, Egyptians, and the Chinese as a medicine with anaesthetic qualities, its popularity grew significantly during the 18th century through recreational use. It was made illegal numerous times in China by successive Chinese emperors. However, the addictive nature of the drug drove demand, in addition to smugglers and corrupt officials seeking profit. By the end of the 19th century, China benefitted from an enormous trade surplus with Europe through their export of porcelain, silk, and tea in exchange for silver. In response, British and American merchants increased their transport of opium to China through British-controlled free ports.[5]

Eventually, this led to a widespread clampdown on opium in 1839, involving the seizure of supplies at Chinese ports and shutting down the factories. After several standoffs and skirmishes, a British military expedition was sent to China in 1840 to seek financial reparations and to guarantee future trade. This resulted in the First Opium War (1839–1842), which resulted in the ceding of Hong Kong to Britain. This was followed by the Second Opium War (1856–1860) between China and western allies (Britain, the US and France), which resulted in the Kowloon Peninsula and Stonecutters Island joining British Hong Kong. The eventual transfer of Hong Kong back to China in 1997 inspired the plot of the James Bond movie *Tomorrow Never Dies* (1997).

Therefore, at the time *The Strange Case of Dr Jekyll* and *Mr Hyde* was published in 1886, the opium trade was legalised once again and was a well-known activity among the public following the Convention of Beijing. The addictive and personality-altering nature of opium is only implied in the early onscreen depictions of *Dr Jekyll* and *Mr Hyde*. Despite knowing the kind of person he becomes as Mr Hyde, Dr Jekyll continues to take his chemical concoction over and over again. It is not until the 1960s that the addictive nature of recreational drugs becomes explicit onscreen, with explicit comparisons made to opium in *The Two Faces of Dr Jekyll* (1960). Because opium was a plant extract, not a synthetic chemical produced in a lab, the connection between professional chemists and opium was not immediately obvious.

It wasn't until heroin and morphine were isolated from opium in the late 19th century for medicinal purposes that the connection with chemistry slowly began to appear. However, at the time, the discipline of chemistry was only seen as a supplier or support to the medical profession, with university chemists often based in medical departments like that seen in *The Mad Ghoul* (1943) and *The Great Moment* (1944), among others. Morphine and heroin were introduced into medicinal use at a time when tuberculosis was rampant, offering some relief for patients. They also found extensive use as pain relief for soldiers during WW1. However, the addictive nature of these opiate drugs eventually became a concern, resulting in restrictions like the US Heroin Act of 1924 among others.

Similar to opium, South Americans chewed the leaves of the coca plant for thousands of years to help elevate mood, aid digestion, and suppress appetite. Until the mid-19th century, the use of coca was only practised in its natural habitat of the Andes mountain range. Coca did not find use in Western medicine until the late 19th century, when American drug companies began to explore that part of the world for new medicines. At first it was considered a safe stimulant and nerve tonic (hence its use in Coca-Cola), but coca's addictive and destructive properties eventually became apparent. When cocaine reappeared in the 1970s as a recreational drug, it was touted as the champagne of drugs because it was expensive, high status, and claimed to have no serious consequences.[6]

In addition to drugs from plant extracts, new synthetic drugs also appeared in the first half of the 20th century, starting with amphetamines in the 1920s and lysergic acid diethylamide (LSD) in 1938. However, widespread availability of these chemicals did not materialise until the industrial expansion of the post-WW2 era in the 1950s and 1960s. The industrial expansion occurred due to an unprecedented population growth, coupled with rapid technological advancement and repurposing of munitions factories for domestic products in the aftermath of WW2.[7]

It was this rapid expansion which resulted in the lax safety standards. This was in addition to a misunderstanding of the dangers associated with scaling up chemistry processes for manufacturing, *i.e.* just because something works on a small scale in a lab, doesn't mean it will work in the same way in large-scale manufacturing. Eventually this led to dozens of industrial accidents throughout the 1970s and 1980s. These accidents created a legacy for the public perception of chemistry and greatly influenced onscreen chemistry at the time. By the mid 1990s studying chemistry was only useful for a 'hair analyst', as mentioned in *Beetlejuice* (1988), and dismissed as no better than playing with a children's chemistry set in *Bad Boys* (1995). Only recently has this image changed, slowly moving onscreen chemistry away from the themes of 'recreational drugs', 'side effects' and 'industrial accidents' which dominated the big and small screen.

13.3 NOT JUST CHEMISTRY

Of course, the chemical sciences aren't the only scientific profession represented onscreen. Throughout this book there have been numerous mentions of other sciences such as biology, physics, and geology, to name just a few. Just like chemistry, the representation of all scientific disciplines suffers from similar inconsistencies, exaggerations, and inaccuracies. During a TV interview, the well-known physicist Neil deGrasse Tyson stated that *Moonfall* (2022) violated more laws of physics per minute than any other sci-fi movie he had ever seen, surpassing what he regarded as the previous record, *Armageddon* (1998). Similarly, geoscientists have no doubt cringed throughout *The Core* (2003) and there are plenty of biologists (and chemists) who have laughed at the use of a micropipette to inject someone in *Maze Runner: The Death Cure* (2018).

There are also numerous examples of movies featuring non-descript scientists who are associated with fictional terminology like those seen in the *Ghostbusters* (1984–present) franchise. Their "proton packs" use "power cells" which are said to have a "half-life of 5000 years". This is perhaps referencing the radio-isotope carbon-14, which is used for carbon dating and has the same half-life. Also, while investigating the properties of the slime in the first *Ghostbusters* movie, they shout random scientific statements at it because it is described as a "psychoactive substance", like "you're nothing but an unstable short-changed molecule" and "you have a weak electrochemical bond". They also perform a "kinetic test" by placing the slime in a toaster, causing it to bang and dance around the place.

Movies like *Ghostbusters* (1984) are undeniably entertaining. Bending the laws of physics, using non-sensical scientific terminology, and exaggerating or simplifying lab techniques probably won't hurt the image of science in any major way overall. It may contribute to false impressions, but as long as these depictions are clearly set in a sci-fi universe like that of the Ghostbusters franchise, we can suspend our disbelief for the sake of entertainment. The main problem is the depiction of 'mad scientists', 'evil scientists', illegal activities and accidents causing harm, since these can lead to negative associations. Overall, chemistry is the scientific discipline most often associated with industrial accidents, as well as illegal activities like the processing of recreational drugs.

In 2015, the Royal Society of Chemistry's Public Attitudes to Chemistry report found that most people were unable to see chemistry as being personally relevant, lacking examples of concrete applications. A quarter felt that school turned them off chemistry and nearly half felt that the chemistry learned at school wasn't useful to them. However, three quarters felt that chemistry had a positive impact on people's lives, in general, but the majority still associated chemists with the development of drugs and medicines. Chemistry was also seen to lack some of the 'fun and energy' of other sciences; being seen as overly serious in personality, and comprising difficult, repetitive experiments.[8]

The influence of movies and TV on opinions and attitudes is difficult to measure. However, it is often claimed that *Top Gun*

(1986) lead to a 500% increase in US Naval recruitment. Recent fact checking has found that there was indeed an increase in the years following the movie, but the number was actually 8%.[9] There are too many variables to blame onscreen chemistry for the negative public opinion of chemistry and/or chemicals, but it is likely that it has contributed. In most cases though, both fictional and non-fictional chemistry depictions were inspired by what happened in the real world. Directly or indirectly, industrial accidents, coverups, illegal activities, new discoveries, problem-solving, and mistakes all contributed to depictions of chemistry on screen. What is very important to note though, is that, despite the evolution of onscreen chemistry across the decades, the image of chemists onscreen is nearly always positive. There are only a handful of examples of truly evil chemists shown onscreen, *e.g.* in *The Mad Ghoul* (1943). In most cases the chemists are depicted as heroes, like in *The Adventures of Tartu* (1943) and *The Rock* (1993), or as problem solvers like in *Lorenzo's Oil* (1990) and *The Martian* (2015) among many others.

13.4 EVOLUTION

So far, we have explored how the overall story of onscreen chemistry has evolved and changed over time. Here we will look at the content, visuals, and language. In much of the early gothic horror depictions there was very little in the way of chemicals mentioned. Also, the lab glassware was mainly used a prop with little or no interaction. However, by the 1930s we start to see several lab skills demonstrated onscreen. Highlights include the unique first-person perspective using common glassware items shown in *Dr Jekyll* and *Mr Hyde* (1931) and a chemist accounting for parallax while pouring a liquid into a measuring cylinder in *The Love Test* (1935). However, the 1930s also saw the beginning of the elaborate glassware apparatus joined together with lots of nonsensical tubing. This elaborate depiction of glassware has stood the test of time and still appears today, *e.g.* in *The Dropout* (2022).

In addition to comparing the plethora of *Dr Jekyll* and *Mr Hyde* adaptations, one of the easiest ways of examining the evolution of chemistry content onscreen, is to compare remakes and re-boots like *The Nutty Professor* (1963) and *The Nutty Professor*

(1996). The latter tones down the colourful liquids in the elaborate glassware apparatus and adds a computer to the familiar transformation scene. A similar comparison can be made between *The Absent-Minded Professor* (1961) and its 1990s remake *Flubber* (1997). Despite the large time span between these two movies, they are strikingly similar, except for the addition of computers and an AI personal assistant robot in the latter. It is noteworthy that the "metastable sphere" shown onscreen in *Flubber* (1997) is an accurate C_{60} buckminsterfullerene (bucky ball) molecule, which was the subject of the Nobel Prize in Chemistry a year earlier in 1996.

It should also be noted that the protagonists in *The Absent-Minded Professor* (1961) and *Flubber* (1997) are both based on a real chemistry professor at Princeton University. He was nicknamed Dr Boom due to his explosive demonstration lectures, providing us with another example of how fiction has been inspired by the real world. Unfortunately, though, the goofiness and silliness of these movies is prioritised above the need to include any sort of real or correct chemistry. But this wasn't always the case for movies with fictional chemistry. Even the 1950s monster movies correctly explained which radioactive isotope they used as a source of gamma radiation, *e.g.* cobalt-60. This is contrasted with *The Absent-Minded Professor* (1961) and *The Hulk* (2003), which only use the term 'gamma rays' with no details provided about the source. As discussed in relation to *Monkey Business* (1952), comedy can be achieved through accurate and informative chemistry without the need for tropes like 'mad scientists', explosions, and incorrect chemistry.

We can also examine onscreen examples with similar themes across different time-periods. For instance, the level of chemistry provided in *The Martian* (2015) is far more detailed then that seen in *Apollo 13* (1995). A similar comparison can be made between *Erin Brockovich* (2000) and *Dark Waters* (2019). Both of these make use of a 'parachute scientist' to briefly explain the chemistry, but the latter is much more detailed and chemically correct. Notably, the evolution of scientific accuracy and detail on screen is not just confined to chemistry. The immunology and microbiology depicted in *Contagion* (2011) is a significant improvement in terms of accuracy over the ridiculous use of monkey plasma as a cure in *Outbreak* (1995).

We can also examine onscreen examples for changes in the use of scientific language. Molecular structures, chemical formulae and chemical equations are core to the language of chemistry. One of the first images shown of a molecular structure onscreen is in *The Adventures of Tartu* (1943), but the grainy picture and hand-drawn notes make it difficult to fully decipher. The next time we see one is very briefly through an image in a textbook in *The Man in the White Suit* (1951), which resembles a type of polymer and appears to be accurate. Correct chemical equations are also shown on the chalkboard in *It Happens Every Spring* (1949) and in *The Affairs of Dobie Gillis* (1953). But the first clear use of editing a molecular structure by a chemist onscreen is in *Lover Come Back* (1961). Overall, prior to the 1970s, most of the chemical structures, formulae and equations were accurate with only some minor errors. This includes, for the most part, the correct use of subscript.

But then, impossible chemical structures appear with the full confidence of showing them on screen long enough to turn the stomach of every chemist who sees them, *e.g. Moonraker* (1979) and *Medicine Man* (1993). There are also several examples of incorrect subscript use for chemical formulae, most notably seen in *Terminator: The Sarah Connor Chronicles* (2008–2009) and *Sherlock* (2010–2017). Some familiar terms have also become vaguer over time, *e.g.* 'truth serum'. In early movies it was nearly always referred to by its chemical name, sodium thiopental, or by its brand name, Pentothal. In both *True Lies* (1994) and *Meet the Fockers* (2004), it is only ever referred to as "truth serum". There are some exceptions through, like *Kill Bill: Volume 2* (2004), in which "truth serum" is used to describe a new substance which is "twice as potent as sodium pentothal". We are also very briefly shown a correct molecular structure in *Good Will Hunting* (1997). The use of molecular structures onscreen then increases in line with the popularity of onscreen forensics. As discussed, some are correct, but most break nearly every law of chemistry. In general, though, accuracy in terms of the language of chemistry has improved immensely since the early 2010s, with correct molecular structures even making their way into animations like *Big Hero 6* (2014).

But what about the 'mad scientists'? As discussed at the beginning of this book, the early gothic horror movies contributed greatly to the established image of the 'mad scientist', but no singular example is fully responsible. Much like Frankenstein's monster, the established image of the 'mad scientist' is actually a combination of body parts, plucked from dozens of examples and cemented during the 1980s. From the very beginning, all sorts of scientists have been called 'mad', 'insane', and 'cracked in the head'. Sometimes this is due to an unforeseen side effect from their discoveries, but it is also related to a difficulty in understanding what motivates scientists to go without food and human contact in favour of persistence and obsession with their work. Eventually, Doc Brown from the *Back to the Future* (1985–1990) trilogy completes the full package of the 'mad scientist' as we now recognise it, with crazy white hair, yellow gloves, lab coat and round safety goggles. But why is the image of the 'mad scientist' so often associated with the early depictions of Dr Victor Frankenstein?

The answer may be as simple as the fact that the comedy horror *Young Frankenstein* (1974) was shot in black and white rather than in colour, as a homage to the original Frankenstein movies. However, Gene Wilder's comically manic depiction, complete with wild hair and round safety googles, has little in common with the early depictions of Dr Victor Frankenstein, who is presented as a young man with neat black hair. This is also evident in the colourised version of *Frankenstein* (1931), which the two protagonists draw inspiration from to create the woman of their fantasies in *Weird Science* (1985). Over time, still images from the 1974 movie may have been mistaken for the 1931 version. As a result, this could be an example of the 'Mandela effect', which is a popularised phenomenon in which "a group of people collectively misremember facts, events, or other details in a consistent manner". It refers to a widespread but false memory that Nelson Mandela died in prison in the 1980s.[10] Like the early photographers who manually transcribed the scenes shown to them by the camera obscura, this could also be a collection of different interpretations.

Overall, the accuracy and level of detail in onscreen chemistry (and other sciences) has improved immensely since the early 2010s, notwithstanding some exceptions. It is proposed here that this is most likely due to the *'Breaking Bad'* effect, *i.e.* the accuracy of the chemistry portrayed onscreen in the *Breaking Bad*

(2008–2013) series encouraged subsequent movies and TV series makers to do the same. The popularity of the *Breaking Bad* franchise, which includes *Better Call Saul* (2015–2022), across a diverse audience demonstrated for the first time that chemicals, complicated names, formulae, and molecular structures do not disengage people. Although the linking of chemists and indeed chemistry teachers with the concept of illegal drug production has created a new (unwanted) stereotype, *Breaking Bad* (2008–2013) finally broke the association of chemistry with industrial accidents.

One of the most prominent themes discussed throughout this book is that of the 'evil industry' or 'greedy corporation'. This theme has not changed or improved throughout the history of onscreen chemistry. Although it is mostly associated with pharmaceutical and chemical companies, it should be noted that it is not just confined to chemistry. It can also refer to biological companies like that seen in *Mission Impossible II* (2000), which features an evil scientific corporation who produce a virus as well as the cure. The aim of the company is to spread the virus, which would provide them with enormous revenue as the only provider of the cure. Similarly, the unethical company at the heart of the Robocop franchise, Omni Consumer Products (OCP), are profit-driven and have their fingers in many pies.

There is no doubt that some chemistry-based corporations have been responsible for terrible and horrific accidents in the real world, usually through negligence like that seen in *The Railway Men* (2023). However, there are very few onscreen examples of the good that most companies have done for humanity, with *Medicine Man* (1992) being the most prominent. From insulin, to vaccines, to modern consumer products, the chemical and pharmaceutical industry have brought important innovations to the masses. It is disingenuous to the research, innovation, engineering, and manufacturing staff in the industrial sector that onscreen they are nearly always associated with accidents, side effects, and greed.

13.5 REPRESENTATION

It may have become obvious as we progressed through the history of onscreen chemistry, but most of the onscreen chemists that we have encountered have been white males. We previously noted some depictions of black chemists in *The Schemers* (1922), which is now considered a lost film, as well as *Miracle in Harlem* (1948) and *Dr Black, Mr Hyde* (1976). But it took another quarter of a century before another black chemist appeared on screen in the form of Samuel Jackson in *The 51st State/Formula 51* (2001). More recently *Black Panther* (2018) and *Black Panther: Wakanda Forever* (2022) have finally given us the first fictional female black scientist on screen who occasionally does chemistry.

We also mentioned examples of chemists who are implied to be Asian like *Charlie Chan at the Race Track* (1936), but the actor (Warner Oland) is white with no proven Asian background. Better examples include *Godzilla* (1954), the *Super Giant* (1957–1959) film series and *Dogora* (1964), among others. Like English-language movies, themes such as invisibility appear in the Tamil-language *Maya Manithan* (1958), 'mad scientists' appear in the Hindi-language *Shikari* (1963), a version of the *Dr Jekyll* and *Mr Hyde* story is told in the Malayalam-language *Karutha Rathrikal* (1967), and a drug to bring the dead back to life appears in the Tamil-language *Nalaya Manithan* (1989). More recently we have well-established chemistry themes such as poisoning and school labs in the Malayalam-language horror *Chemistry* (2009) and depictions of Indian chemists in the true story of the Bhopal disaster in *The Railway Men* (2023). It should also be noted that *The Railway Men* (2023) is another example of the evolution of onscreen chemistry, since *Bhopal Express* (2001), the previous screen adaption of the Bhopal disaster, only used the word 'gas' with no further details provided.

Turning to female representation, we began this book with a depiction of a female chemist in a lead role in *Irrungen/Mistakes* (1913). The inter-war period then gave us several good depictions which include *The Love Test* (1935), *Beauty for the Asking* (1939), *Kid Glove Killer* (1942) and *Strange Impersonation* (1943). It could also be argued that the witch in *Snow White and the Seven Dwarfs* (1937) is a chemist of sorts, since she makes the poison apple using recognisable lab glassware. We also have the first biography of Marie Curie in *Madame Curie* (1943), depicting Marie juggling her professional and personal life.

Then, after WW2, female chemists disappeared from screen, and it would take a quarter of a century before we meet one again in *Caprice* (1967). However, although she is the inventor of the titular product and runs the lab in which it is made, she only appears in a brief scene. The next example is from the harrowing true story of *Silkwood* (1983), in which the protagonist is a lab technician who manipulates radioactive material. There is then a fourteen year wait for another depiction of a female fictional chemist in a lead role, with Elisabeth Shue playing an electro-chemist in *The Saint* (1997). Note that Elisabeth Shue also portrays a research scientist who does chemistry in *Hollow Man* (2000). Also worth a brief mention here are several scenes with Laura Dern playing the role of a science teacher in *October Sky* (1999). It might be argued that Scully in *The X-Files* (1993–2002) is an example, but she normally identifies as a medical doctor, not a chemist.

In contrast, there are several examples of female biologists onscreen during this period, such as Kate Reid as a microbiologist in *The Andromeda Strain* (1971), Sigourney Weaver as a primatologist in *Gorillas in the Mist* (1988), Laura Dern as a palaeobotanist in *Jurassic Park* (1993), Reno Russo as a CDC scientist in *Outbreak* (1995), Julianne Moore as a clumsy epidemiologist in *Evolution* (2001), and Kate Winslet as an accomplished epidemiologist in *Contagion* (2011). The small screen also gave us Emily Deschanel as a "forensic anthropologist" in the lead role of *Bones* (2005–2017) and there is a female microbiologist and a female neuroscientist represented in *The Big Bang Theory* (2007–2019). It's also a similar story for female physicists and mathematicians on screen, which include *Chain Reaction* (1997), *Contact* (1997), *Interstellar* (2014), *Hidden Figures* (2016) and *The Big Bang Theory* (2007–2019). Of course, this is not to say that these depictions are flawless, since they often contain outdated tropes of female scientists like the 'daughter/assistant' or the 'male woman' among others. But they significantly outnumber female chemists onscreen.

The turn of the millennium provides an increase in the number of female chemists depicted onscreen, usually in association with TV forensics. However, many of these depictions are in connection with incorrect or exaggerated chemistry and the forensic chemist is usually a secondary character. In recent years, the only depictions of female chemists in lead roles are biographies about real people like Marie Curie in *Radioactive* (2019) or *Clara Immerwahr* (2014). Clara was the daughter of a chemical industrialist and became the first German woman to be awarded a doctorate in chemistry from the University of Breslau. Later, she also married the chemistry Nobel Prize winner Fritz Haber, as mentioned previously.

Clara also spoke out against her husband's research into chemical weapons, calling it a 'perversion of the ideals of science'. *Clara Immerwahr* (2014) shows several scenes in labs with chemicals, glassware, and various equipment. It also includes scenes in chemistry teaching labs with the correct chemical formula for formaldehyde written on the chalkboard. There are also some experiments shown and discussed including the well-known demonstration 'thunderstorm in a tube', involving potassium permanganate, sulfuric acid, and ethanol. Several lab skills are also presented, including wafting a smell from a container. Later in the movie, there is also an explosion in Haber's lab which alerts Clara to the new work of her husband in the area of poison gas, as discussed here previously.

But Marie Curie and Clara Immerwahr are real people – what about fictional female chemists? *Lessons in Chemistry* (2023) represents the first fictional female chemist onscreen in nearly a quarter of a century and is based on the popular book of the same name by Bonnie Garmus. Set in the 1950s, it follows a chemist named Elizabeth Zott (Brie Larson) who can no longer practise chemistry because she is pregnant and unwed. The miniseries opens with Elizabeth entering a television studio to run her cooking show *Supper at Six* and asking if they have "the sodium chloride I requested". She then dismisses a brand of soup because "it's full of chemicals, and not the good kind". It then flashes back to her time at a research institute which provides us with most of the onscreen chemistry, but there are references to food chemistry scattered throughout the series.

The lab at the Hastings Research Institute opens with plenty of empty lab glassware scattered around the benches, with Elizabeth acting as the lab tech. Subsequent scenes also show condensers attached to large boiling flasks set up for reflux, but none of them contain colourful liquids or random tubing. As a result, the labs are some of the most realistic seen onscreen in many decades. There is also a familiar reference to the theme of funding, when the famous male chemist Calvin (Lewis Pullman) is told that the grants "keep the lights on around here". Initially he is presented as obsessive and eccentric, but he also suffers from various allergies.

Elizabeth is presented as knowledgeable in chemistry, with the male chemists in the lab asking her for advice. She tells one "you're not seeing conversion of the carbonyl group into the amine group? You forgot to add the acid catalyst". This is a real chemical reaction known as a reductive amination, and the reaction conditions are usually neutral or weakly acidic, but a reducing agent is normally used as a catalyst. She also makes coffee in the lab using lab glassware and a Bunsen burner, but she doesn't wear safety goggles. In fact, she doesn't wear safety goggles at any stage during the series. In a throwback to *The Love Test* (1935), the male staff are frequently sexist and dismissive of Elizabeth.

It then transpires that she is moonlighting in the lab to conduct her own biochemical research, but she is caught and reported. This is when we find out that she has a Master's in chemistry, during which she studied the "cellular metabolism of nucleic acids". But she is told that she "is not smart enough" for the research grant and she should instead enter the beauty contest. Later, when talking to the other beauty contestants, she tells them that she doesn't wear perfume because her sense of smell is compromised after an incident with hydrochloric acid. As she tells this story, the faces of the other contestants glaze over, establishing that she stands apart from them.

There are strong hints at elitism throughout the series, since only those with PhDs are considered to be chemists. Her frequent claims to be called a chemist are dismissed, despite her qualifications. She then befriends Calvin and experiments with different recipes as they have lunch together while discussing biochemistry problems. This provides us with several references to the story of Marie and Pierre as told in *Madame Curie* (1943). First, when Calvin offers her some research space in his lab, they argue over which part of the lab is theirs. Secondly, Calvin's absent-mindedness is compared to Pierre through the same foreshadowing technique, *i.e.* in the first episode Calvin nearly gets hit by a bus when crossing the road without looking, which is a homage to *Madame Curie* (1943) where Pierre nearly gets hit by a horse and cart at the beginning of the movie. Thirdly, through Elizabeth's determination to continue their work against all odds.

Unfortunately, though, their lab includes an example of anachronism (time error) when multiple rotary evaporators are shown. It is stated during the beauty contest that the year is 1951. Although it was invented in 1950, this common lab device was not commercialised until 1957 and didn't become common place until the early 1960s.[11] Commendable though are the molecular structures shown on a chalkboard in the lab, one of which is the correct structure of ferrocene which was discovered in 1951.[12] The correct structure of formaldehyde is also shown, as well as a pentavalent phosphorus structure with no apparent chemical errors. Elizabeth then corrects this structure into a dioxolane type structure, which again has no chemistry errors (Figure 13.2).

Worth a brief mention also is a joking reference to cyanide, which is interesting because in the same scene we also learn that Calvin is allergic to benzaldehyde. Cyanide and benzaldehyde both smell like almonds. Benzaldehyde can cause a skin allergy in real life, but probably not vomiting as shown. Also noteworthy is when Elizabeth mentions another famous female chemist, Dorothy Crowfoot Hodgkin. Overall, *Lessons in Chemistry* (2023) is a welcome return to realistic labs and chemistry in a fictional environment, with Brie Larson doing an expert job at discussing chemistry and performing several laboratory skills accurately. Also noteworthy is the reference to 'evil industry' in the form of the research institute who, through an executive, engages in

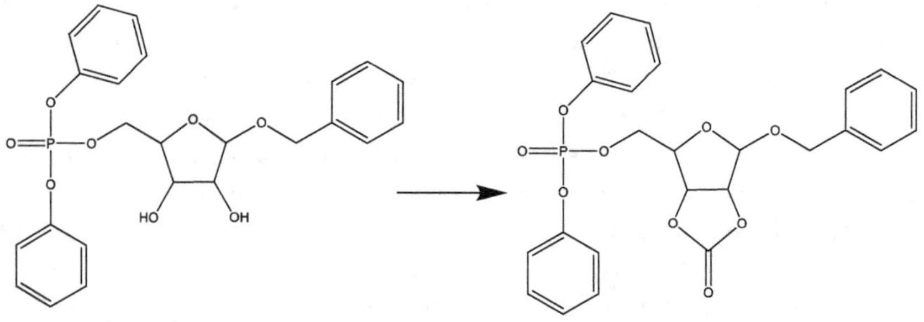

Figure 13.2   The molecular structure drawn by Calvin (left) and corrected by Elizabeth (right) as shown in Lessons in Chemistry (2023).

unethical behaviour. Finally, the series ends with a chemistry lecture by Elizabeth, again referencing *Madame Curie* (1943).

Lessons in Chemistry (2023) is also a welcome return to the theme of 'food chemistry' and serves as a vast improvement in terms of the onscreen chemistry compared to all the movies featuring Willy Wonka, Charlie, and the chocolate factory. Noteworthy also is a brief mention of chemistry in the food-based comedy drama *The Bear* (2022–present), when sodium bicarbonate is mentioned during their fire suppression test. Sodium bicarbonate is used as a real fire suppressant, providing thermal cooling and, when heated, it produces water and carbon dioxide which starve the fire of oxygen.[13] So perhaps this is the beginning of a new theme for onscreen chemistry, but only time will tell.

13.6 CONCLUSIONS

During the 19th century, chemists generally enjoyed public support because of their numerous advances in medicine, nutrition, and materials. However, public perceptions of chemistry eroded during the early half of the 20th century mainly due to its associations with chemical warfare. The post-WW2 era offered some reprieve when the further advancements provided by chemistry research and industrialisation once again made people's lives better. But this was followed by a decline in popularity during the accident-prone era of the 1970s and 1980s. By the time we reached the 1990s, chemistry depictions became so brief that they we were reduced to the choice between a red pill and a blue pill in *The Matrix* (1999), echoing the story of *Alice in Wonderland*. Of course, we are never told what the pills contain.

It is unclear whether the association of *Dr Jekyll* and *Mr Hyde* with the duality of chemistry was purposeful or not. From the numerous examples discussed in this book, it would seem that this association was on purpose, due to several explicit links that we have seen. One of the most prominent of these links between the gothic story and chemistry is the use of mirrors in some screen adaptions. As discussed previously, when Jekyll looks in the mirror and sees Hyde, this is very likely a reference to chiral molecules. In chemistry, a molecule is called 'chiral' when it cannot be superimposed on its mirror image by any combination of rotations or translations. The molecules have the same chemical formula, but their molecular structures are said to be 'non-superimposable mirror images' or 'enantiomers' of each other. This can result in different chemical properties. One of the best real-world examples is the artificial sweetener aspartame, which has two enantiomers. We use L-aspartame in food because it tastes sweat, whereas D-aspartame is tasteless. Understanding chirality is also very important for drug safety and effectiveness, with one version providing beneficial effects and the other being toxic, *e.g.* D/L-penicillamine. So even within chemistry itself, there is duality.

The Royal Society of Chemistry Public Attitudes to Chemistry survey in 2015 also found that most people associated chemistry with school and medicine, which should come as no surprise considering the onscreen chemistry examples of the past three decades. We have seen through numerous examples how the early screen adaptations of *Dr Jekyll* and *Mr Hyde* influenced later onscreen chemistry examples. These early gothic horror depictions accurately presented chemistry with a sense of duality, which has remained relevant throughout the entire history of onscreen chemistry, for better or worse. But this duality was compromised for a period from the early 1970s to the early 1990s, with a shift towards the negative aspects. It is proposed here that the numerous depictions of chemistry-related industrial accidents and illegal activities shown onscreen, both fictional and non-fictional, influenced public perceptions of chemistry and chemicals. During this period, chemistry faced warranted criticisms and it is only since the 1990s that the duality has re-emerged with a positive focus on education, medicine and, more recently, food chemistry.

Also, since the early 2010s, there has been a noticeable improvement in the number and accuracy of chemistry depictions. The most recent examples indicate that we might be entering a new golden era for onscreen chemistry, which includes younger audiences in *Elemental* (2023) and female representation in *Lessons in Chemistry* (2023). As a result, onscreen chemistry may be returning to its creative and problem-solving roots. However, have all the consequences of this been considered? The human and environmental costs of modern society are the cross that the chemistry community must continue to bear, so horrific true stories like *The Railway Men* (2023) will also continue to appear. But, overall, we can take solace in the fact that onscreen chemists are usually depicted on the side of good and they generally want to make the world a better place.

REFERENCES

1. C. Harman, *Robert Louis Stevenson*, Harper Perennial, 2010.
2. R. L. Stevenson, *The Strange Case of Dr Jekyll and Mr Hyde and Other Stories*, Alma Classics, 2014.
3. P. D. Mitchell, Human Parasites in the Roman World: Health Consequences of Conquering an Empire, *Parasitology*, 2017, **144**(1), 48–58.
4. K. Olsen, Clear Waters and a Green Gas: A History of Chlorine as a Swimming Pool Sanitizer in the United States, *Bull. Hist. Chem.*, 2007, **32**(2), 129–140.
5. K. Pletcher, *Opium Wars*, Encyclopædia Britannica, Inc., https://www.britannica.com/topic/Opium-Wars.
6. A. López-Valverde, J. De Vicente and A. Cutando, The Surgeons Halsted and Hall, Cocaine and the Discovery of Dental Anaesthesia by Nerve Blocking, *Br. Dent. J.*, 2011, **211**(10), 485–487.
7. C. Bohannon, Economic Recovery: Lessons from the Post-World War II Period, Mercatus Centre, George Mason University, 2012, https://www.mercatus.org/research/policy-briefs/economic-recovery-lessons-post-world-war-ii-period.
8. E. Fu, A. Fitzpatrick, C. Connors, D. Clay, B. Toombs, A. Busby and C. O'Driscoll, *Public Attitudes to Chemistry*, Royal Society of Chemistry, 2015, ch. 1, pp. 19–39.
9. W. Summers, Maverick Top Gun stat turns out to be a real goose, AAP Factcheck, https://www.aap.com.au/factcheck/maverick-top-gun-stat-turns-out-to-be-a-real-goose/.
10. M. McDonough, *Mandela Effect*, Encyclopædia Britannica, Inc., https://www.britannica.com/science/Mandela-effect.
11. W. B. Jensen, The Origin of the Rotavap, *J. Chem. Educ.*, 2008, **85**(11), 1481.
12. G. B. Kauffman, The Discovery of Ferrocene, the First Sandwich Compound, *J. Chem. Educ.*, 1983, **60**(3), 185–186.
13. A. E. Finnerty, Fire-Extinguishing Powders, *Halon Options Technical Working Conference*, May 1997, pp. 206–215.

Subject Index

Page numbers in *italic* refer to figures.